Praise for *Blue Frontier*

"[Helvarg] takes us on the ultimate wave, cresting and carrying us at breakneck speed. It's a fast, watery ride, and you're going to get wet before it's over."
—*Los Angeles Times Book Review*

"A vivid tapestry of an America interwoven with the sea."
—*San Jose Mercury News*

"David Helvarg takes us from coast to coast and below the sea in this illuminating, insightful, and sobering look at our imperiled oceans—and the challenges we must overcome if we are to save our 'blue frontier.'"
—Ted Danson, founding president, American Oceans Campaign

"Reads like a scientific detective novel."
—*Sacramento Bee*

"David Helvarg's insightful new book underscores the full measure of the challenge before us: If we hope to explore the blue frontier, we must travel cautiously, repairing the damage we have done, understanding before we exploit, and always preserving the natural systems that have created it."
—Senator John F. Kerry

"A compelling case for redefining this country's ocean policy."
—*San Diego Magazine*

"Helvarg weaves history, marine biology, and politics into an informative, readable account of the struggle to save our ocean environment. A must-read for anyone seeking to put the current debate in context."
—Jim Toomey, syndicated cartoonist, *Sherman's Lagoon*

"David Helvarg has created an infuriating portrayal of mankind's most breathtaking crime, the destruction of the oceans that birthed our species."
—Robert F. Kennedy, Jr., president, Water Keeper Alliance

"An essential addition to the environmental science library."
—*Choice*

"This whirlwind tour highlights the intertwining challenges facing America's coasts and oceans. Helvarg's on-site and at-sea narrative puts you where the action is."

—Carl Safina, author of *Song for the Blue Ocean*

"I thought I knew the ocean, until I read David Helvarg's book. *Blue Frontier* is a fascinating read about the importance of our seas to our quality of life and our future."

—Alexandra Paul, actress, *Baywatch*

"Two things dazzled me about this book. First, author David Helvarg's research is amazing. Second, the way he presents his information is fascinating."

—*Toledo Blade*

Blue Frontier

Saving America's Living Seas

DAVID HELVARG

A W.H. Freeman / Owl Book
Henry Holt and Company
New York

Henry Holt and Company, LLC
Publishers since 1866
115 West 18th Street
New York, New York 10011

Henry Holt® is a registered trademark of
Henry Holt and Company, LLC.

Library of Congress Cataloging-in-Publication Data

Helvarg, David, 1951–
   Blue frontier : saving America's living seas / David Helvarg.
        p.   cm.
   Includes bibliographical references and index.
   ISBN 0-8050-7135-0 (pbk.)
   1. Marine resources conservation—United States. I. Title.
   GC1020 .H45 2001
   333.91'6416'0973—dc21   00–010956

Henry Holt books are available for special promotions and premiums.
For details contact: Director, Special Markets.

First published in hardcover in 2001 by W. H. Freeman and Company

First Owl Books Edition 2002

A W. H. Freeman / Owl Book

Designed by Blake Logan

Printed in the United States of America

1 3 5 7 9 10 8 6 4 2

To Ed Ricketts
For making it fun,

To Rachel Carson
For celebrating the life,

To Roger Revelle
For having some regrets,

And to Rell Sunn
For surfing through the pain.

Contents

I really don't know why it is that all of us are so committed to the sea, except I think it's because in addition to the fact that the sea changes, and the light changes, and ships change, it's because we all came from the sea.

—PRESIDENT JOHN F. KENNEDY, 1962

Blue Frontier

Thrashed

And I have loved thee, Ocean! and my joy
Of youthful sports was on thy breast to be
Borne, like thy bubbles, onward; from a boy
I wantoned with thy breakers. They to me
Were a delight; and if the freshening sea
Made them a terror, 'twas a pleasing fear

—LORD BYRON, 1818

Catch a wave and you're sitting on top of the world

—THE BEACH BOYS, 1963

I turn around and catch Tim's longboard straight across my throat. I go under and come back up and look down Stinson Beach and it's all red; the people, the pine covered hills that were green a moment ago, the sand and breaking waves are all crimson to my eye. I cough a dry hacking cough because it feels like a large twig's just snapped inside my windpipe. Tim asks if I'm okay. "Everything's red," I croak, but even as I do so, the blood begins to recede from behind my eyeballs. He grabs the 10-foot surfboard I'd been trying out before it got loose in the waves and offers me his free arm, but I wave it off, staggering ashore. We strip off our wetsuits back at his car. I've gone hoarse and raspy but the karate chop pain's now receded to a constant scratchy constriction of my throat. We drive back to my place in Sausalito and later I go to the hospital where I get tubed in a CAT scan. The doctor tells me my larynx is bruised but not broken and sends me home with some codeine and steroids. I can't eat for several days, sound like Marge Simpson for two weeks, but other than that I'm

1

fine. Not really fine. I'm about to leave the ocean and the shore that I love for the first time in more than 20 years, heading to Washington, D.C., for work and other reasons. My East Coast cousin, after hearing of my accident, asks, "Aren't you a little old to be surfing?" and it makes me even edgier about the move. A man in his forties, splinting his shins jogging around the Central Park Reservoir in Manhattan or through Washington's Rock Creek Park is considered sensible, taking care of his health, even if he gets shredded by a Doberman. But find your pleasure in the ocean off northern California where a few big sharks feed, and you're suspected of being an arrested adolescent, of taking needless risks. Even those I've come to know who earn their living from the sea—sailors, fishermen, oil platform roughnecks, commercial divers, charter boat operators, and lifeguards—are often accused by the mainstream culture of "running away from life," of not facing up to the responsibilities that come with low-risk, highly regimented work far from the siren song of the everlasting sea.

Actually it's the same accusation faced by an earlier generation of greenhorn adventurers who left the fetid urban ghettos of the eastern seaboard to settle the arid open spaces of our nineteenth-century frontier. Frontiers I've discovered in visits to places like coastal Alaska and maritime Antarctica tend to attract fools and visionaries. And like those earlier frontier settlers who clashed with the native populations and slaughtered the indigenous wildlife, today's ocean settlers—be they beachfront home owners on exposed barrier islands, drag trawl fishermen tearing up benthic communities of bottom-dwelling life, or recreational divers drifting with turtles and sharks—are facing many of the same problems and perils seen on America's last great wilderness range.

Back in 1890, just a year after the Oklahoma Land Rush, the U.S. Census Bureau ended a key chapter in American history by declaring the nation's frontier closed. But on March 10, 1983, President Ronald Reagan, in one of the most significant and least noted acts of his administration, opened up 3.4 million square nautical miles of new territory, extending U.S. sovereignty over a wet frontier six times the size of the Louisiana Purchase, 30 percent larger than the continental land base of the United States. It's an oceanic domain that stretches from New England's Georges Bank to St. Croix, Virgin Islands, and from Dutch Harbor, Alaska, to beyond the outer reefs of Guam. But unlike our last frontier, the creation of this new blue one, our Exclusive Economic Zone (or EEZ), has failed to spark the public imagination, to inspire grand plans and visions, or even to resolve the ongoing competition and struggle over our nation's maritime resources. That conflict, however, could lead either to the protection and sustainable use of America's greatest natural treasure or condemn our oceans to a final industrial onslaught of destruction, leaving them fit only for the sulfur-eating seaworms and giant bacterial clams of Oregon's deep-ocean geothermal vents and chimneys, assuming these mineral-rich areas are not themselves strip-mined by high-tech dredges and robots.

But it's best we start with the given and the known about this, our final physical frontier. The seas cover 71 percent of the earth's surface, giving our ocean planet its blue marble appearance. Although the tropical rain forests have been called the lungs of the world, the oceans actually absorb far greater amounts of carbon dioxide. Microscopic phytoplankton in the top layer of the sea act as a biological pump, extracting some 2.5 billion tons of organic carbon out of the atmosphere annually (replacing it with 70 percent of the life-giving oxygen we need to survive). The top 2 feet of seawater contain as much heat as the entire atmosphere. Scientists who recently have come to recognize ocean currents as key to the creation of climate, clouds, and weather still don't know enough about the internal workings of the sea (or have the historic records) to fully incorporate the ocean's thermodynamics into computer models of global warming. More is known about the dark side of the moon than is known about the depths of the oceans.

Photosynthesis of carbon dioxide by plankton and terrestrial plants was thought to be the basis of all organic life until just over 20 years ago, when in 1977 scientists aboard a deep-diving submarine off the Galapágos Islands discovered sulfurous hot-water vents 8,000 feet below the surface of the sea. The area was colonized by giant tube worms, clams, white crabs, and other animals that contain sulfur-burning bacteria that provide an alternative basis for sustaining life. NASA scientists now believe similar "chemosynthetic" life-forms may exist around volcanic deep-ocean vents beneath the icy crust of the Jupiter moon Europa.

If the possiblity of extraterrestrials in our solar system is not enough to excite our celebrity driven culture let's consider the ocean's connection to the O. J. Simpson and Monica Lewinsky cases.

Deep sea vents, chimneys, and black smokers are also home to a number of recently discovered microbes and whole new classifications of microbial species such as *Archaea*, or the "ancient ones"—chemosynthetic life-forms thought to be the first living creatures on earth. Among these is *Pyrococcus*, which acts as the key to the polymerase chain reaction (PCR), the replication engine that drives the DNA "fingerprinting" process that can match DNA strands to individuals with a surety of about one in a billion. It was this type of DNA matching that played such a prominent role in the O. J. Simpson trial, as well as with the infamous presidential otain on Monica Lewinsky's blue dress.

Leaving Planet Hollywood and returning to planet earth, photosynthe-sizing phytoplankton, aside from cleansing our atmosphere, also provide the big pasture at the bottom of the marine food web. For millions of years the ocean has maintained a fecundity of life unmatched on land, an enthralling variety of creatures and wealth of protein that has in the last half century jumped from a 20- to a 90-million-metric-ton annual harvest for human consumption (about 16 percent of the animal protein we consume). This biomass is equal in weight to more than 900 fully armed aircraft carriers being dredged up from the world's oceans every year (as opposed to the dozen U.S. carriers that actually sail the seas). With technologies developed by the military—including radar, sonar, improved navigation and commu-nications systems, satellite surveillance, stronger marine engines, nylon for netting, and strengthened steel and fiberglass hulls—the world's fishing fleets have been waging a highly efficient market-driven war of extermina-tion on a growing list of fish species and marine creatures. As a result, the 1990s and beginning of the twenty-first century have seen a precipitous decline in the world's catch with over 70 percent of commercial fisheries now fully exploited, overexploited, or at risk of collapse, according to the United Nations' Food and Agriculture Organization.

This unsustainable killing occurs even though, along with its practical role in maintaining the tides of life, our ocean planet also holds a spiritual resonance for our species, calling us back to a common waterborne birth state we've all experienced on both an individual and evolutionary basis. Our bodies, like the planet, are 71 percent saltwater; our blood exactly as salty as the sea. This may explain why it's easier to fall asleep to the sound of the ocean. The rhythm of the waves is like our mother's heartbeat.

For seven years I lived with two roommates in a brown clapboard house 60 feet above the pounding Pacific, on an iceplant-covered cliff in San Diego. I never slept better in my life. From our back porch I could watch gray whales migrating in winter, keeping an eye out across the slate-colored waters for the spray clouds of their breath. Sometimes they'd breach, leap-ing full-bodied toward the sky. Sometimes, while out sailing with friends, we'd see them pass within yards of us, their arching backs slick and crusty with barnacles. In the spring and summer, when hollowed-out aquamarine waves left rooster tails of spray in the diamond-dappled sea, I'd scale down the sandstone cliff with Charlie or Manny, green Churchill bodysurfing flippers in hand, and dive off the rocks to catch the breaking wave off Crab Island, a small outcrop of rock just to our north. On more than one occa-sion, on our short amble to the water, we'd encounter a hauled-up sea lion or tidepool octopus trying to blend in with the rocks. But even as we took our pleasures in the sea, we never forgot that we were also living on the edge of a vast and threatening wilderness. Once I had to throw one of Charlie's longboards down to a pair of lifeguards trying to rescue a young girl who'd fallen off the narrow sandstone path that ran below our deck. She died on the way to the hospital. Another time we saw a cabin cruiser breaking up in the surf. Charlie, a cameraman for the local CBS affiliate, grabbed his video camera, while I took hold of the sound deck. By the time we joined the paramedics at the beach below Pescadero Street, it was getting dark. We provided illumination from the camera's portable light as they applied their shock paddles to the victim's chest, a heavy-set man who'd just washed up in swimming trunks and a white T-shirt. Again it was too late. Three of the six people on that boat drowned. It's like Ken Kelton, a kayaker I interviewed who'd been attacked by a great white shark, explained: "The ocean is a dangerous place, but it's also a place you can still go and have to yourself, a place that's clean, and, yes, wild. If you go in the ocean, you're making a choice. You need to know you can drown, you can get lost, or you can be eaten by great beasts."

Now thanks to the late Jacques Cousteau, Emile Gagnan, and the Self-Contained Underwater Breathing Apparatus (scuba) they invented, you can spend quality time with the beasts and the beauties: whale sharks; makos; squid; humpbacks; parrot fish; squirrelfish; frogfish; stonefish; moray eels; turtles; stellar seals; hammerheads; barracudas; starfish; strawberry anemones; sea hares; spiny lobsters; groupers; dolphins; brain, whip, leaf, and staghorn corals; barrel sponges, sea dragons, and rainbow-colored nudibranches. The ocean frontier has a greater biodiversity of life than the richest terrestrial habitats on earth, including rain forests. Disrupting any part of this oceanic ecosystem—the humble spiny urchin as well as the magnificent bluefin tuna—can affect the whole in ways we still don't fully understand. Our actions on land—be it overfertilizing corn fields in Iowa or golf courses in St. Louis, running factory farms in Maryland and Alabama, dripping hydraulic fluid on LA freeways—can (by way of watersheds, rivers, and storm drains) create massive nutrient-fed algal blooms and anoxic (oxygen depleted) dead zones in our coastal waters as has already occurred in the Gulf of Mexico. This is why there is a desperate need to develop and expand not only our biological knowledge of the seas, but also an active and educated political constituency to protect the oceans' living resources.

Unfortunately, our politicians and national leaders seem to be suffering anoxia of the brain when it comes to understanding the value of our living Blue Frontier.

In 1995 the self-styled right-wing "revolutionaries" of the 104th Congress abolished the House Merchant Marine and Fisheries Committee after 107 years of operation. Noted for its bipartisan commitment to marine issues (43 of its last 48 members were from coastal states), some of its oceanic responsibilities were shunted off to a subcommittee of the House Resources Committee. Under the chairmanship of Congressman Don Young of Alaska, this committee was packed with western Republicans from places like Nevada, Utah, Wyoming, and Idaho, who spent more time railing against the reintroduction of wolves into Yellowstone Park than considering the plight of America's vast seas.

Congress also attempted to abolish the Department of Commerce without realizing that its largest division, with some 8,000 people, worked for the National Oceanic and Atmospheric Administration (NOAA), America's lead agency on the Blue Frontier. Only after the U.S. Chamber of Commerce and other business interests pointed out the commercial importance of the U.S. Weather Service, National Hurricane Center, and other NOAA operations did the congressional firebrands retreat.

Three years later, in 1998, the Clinton administration, looking for some way to respond to the UN-declared International Year of the Ocean, sponsored a national ocean conference by the shoreline in Monterey, California. But this turned out to be little more than a scenic photo-op for the upcoming presidential campaign of Vice President Al Gore, with few, if any, new initiatives taken (Clinton did promise to extend a partial off-shore oil-drilling moratorium that was already in effect). Still the conference's title, "Oceans of Commerce, Oceans of Life," at least identified the government's policy priorities. It's not particularly surprising that business should trump biodiversity on America's Blue Frontier when you realize that, since its creation in 1970, NOAA has been anchored deep within the confines of the trade-driven Department of Commerce.

Two years after the oceans conference, candidate Al Gore pledged to ban all oil drilling off the key electoral states of California and Florida (a pledge Florida governor Jeb Bush could not get his brother, Republican presidential candidate George W. Bush, to go along with). Despite much gnashing of teeth and tearing of garments by the oil industry, the reality is that 97 percent of America's future offshore oil potential remains in the deep waters of the Gulf of Mexico where a new, little noted energy boom is occurring.

The years 1997–1999 were also marked by Congress's refusal to pass an "American Oceans Act" that would establish a national blue-ribbon commission to consider the plight of America's Blue Frontier. The American Petroleum Institute, not wanting to jeopardize their position of power in Washington, lobbied hard against the measure on Capitol Hill. Navy officials, worried about new environmental players interfering in their national security projects, also let the administration know they didn't much like the idea. Finally, in 2000, the establishment of an independent oceans commission by the Pew Charitable Trusts, that included fishermen, scientists, two governors, and a former White House chief of staff, helped inspire Congress to pass its own Oceans Act.

On the evening of June 15, 2001, President George W. Bush named 16 members to this new federal commission. The majority were from the offshore oil industry, the navy, and the shipping and ports industry. Notable by their absence were fishermen, marine biologists, and ocean protection activists. The fact that this commission's creation was announced on the White House Web site, along with other Friday night presidential housekeeping duties, reflected the out-of-sight, out-of-mind attitude that continues to define the government's approach to the Blue Frontier.

Still, if the two ocean commissions take an unbiased approach to their work, they're likely find that the fencing off of the American seas, like the fencing off of our prairies more than a century ago, has failed to slow a process of chaotic and rapacious frontier development similar to what took place in the Wild West. Instead of buffalo hunters and cattlemen killing off native animals and replacing them with cows that overgrazed the range and trampled the rivers, we now have giant factory trawlers and draggers overharvesting our seas and destroying bottom habitat even as dams, development, and polluting fish farms threaten the water quality, genetic diversity, and complex ecosystems necessary for wild fish to thrive.

In place of army forts and anti-Indian campaigns we have a post–Cold War navy moving from blue water to brown, looking at the continental shallows and beachfront littoral, the areas where our living resources are most at risk, as their next staging area for war-fighting, extended combat exercises, and coastal bombing practice.

And just as the Seventh Cavalry gave protection to profit-hungry gold miners, opening up Dakota's Black Hills to mineral extraction, today's coastal real-estate developers are being given special dispensations by the Army Corps of Engineers to fill in wetlands and mangrove swamps that act as the nurseries and filters of the sea, undermining laws like the Clean Water Act in their rush to accommodate more high priced "gold coasts."

The silting up and poisoning of salmon and trout rivers near the nineteenth-century frontier mining towns is also magnified ten thousandfold in the 32 billion gallons a day of agricultural, urban, and industrial runoff including oil, pesticides, and manure that are suffocating our coastal bays and estuaries, poisoning marine mammals, and feeding outbreaks of pfiesteria and harmful algal blooms. These threaten human health and safety, while resulting in some 11,000 beach closures a year.

And where once a corrupt Congress sold off the public lands to the railroad trusts for pennies on the dollar, today's Mineral Management Service holds fire-sale leases of off-shore oil and gas while Congress declares royalty holidays and tax breaks for their friends in the offshore industry.

Still, despite today's wide-open frontier activity, the declaration of an EEZ has also provided Americans with an opportunity for a new approach to ocean stewardship, the recognition that when you claim sovereign rights over large parts of the sea you're also taking on a higher level of responsibility for the sea's protection. It's a mission that growing numbers of coastal citizens, local governments, and waterfront communities are no longer waiting patiently for Washington or their statehouses to assume.

They have begun not a grassroots campaign but a seaweed rebellion, and, like the giant kelp plant once it has found a holdfast, this movement has the potential to grow at a terrific rate. It's a rebellion that can be seen from the Web pages of surfer activists ("No way Dude! We don't want your Crude!") to the cleanup, rediscovery, and celebration of historic waterfronts in Portsmouth, Boston, Baltimore, Jacksonville, Galveston, San Diego, Monterey, Seattle, Seward, Hilo, and hundreds of other maritime communities large and small. It's the annual volunteer beach cleanups and diver fish counts, the establishment of marine sanctuary citizen advisory panels, the sailing trips for youths at risk, and marine education programs like the Los Marinaros program for grade-schoolers in Salinas, California, the high school Mast Academy in Miami, Florida, and the Discovery Hall Program for 16,000 students a year at Alabama's Dauphin Island Sea Lab.

It's sometimes angry environmental protests at public hearings on development and offshore oil, and fun but messy restoration projects that can aid the ocean's healing, be it of a muddy duck pond in Rhode Island, a coastal river in Oregon, or the Everglades of south Florida. It's coalitions of fishermen and conservationists, marine scientists, and urban planners in the Northwest, fighting to restore the iconic wild salmon and protect its damaged habitat from rural Idaho to urban Seattle, where the mayor argues that "by saving the salmon we may very well be saving ourselves." It's thousands of outraged letters and newspaper editorials when the former governor of California tried to fire the conservation-oriented director of that state's Coastal Commission, and the "Vote the Coast" coalition that helped elect a more ocean-friendly administration. It's a little girl sleeping snuggled up against her stuffed dolphin or squealing with delight when her father holds her in a wave. It's the future.

For now, it's an uncertain future based on a largely inchoate rebellion, not effectively organized to take the fight to every coastal statehouse, the halls of Congress, or beyond. Still, like green phosphorescence sparkling in a wine-dark sea, it's more than an illusion. It's more like a reconciliation of the soul among tens of millions of Americans both young, and young in spirit, who have come to recognize the limitless possibilities that a living sea has to offer. It's a damp and salty uprising aimed at nothing less than the recovery of our maritime culture and heritage, nothing less than the renewal of our journey home to the American sea.

[Chapter One] Fool's Gold

The deep oceans contain oil and mineral wealth to rival Alaska's North Slope and California's Gold Rush.

—TIME MAGAZINE, 1995

The sea is not a bargain basement.

—JACQUES COUSTEAU, 1963

If the opening of America's western frontier can be traced back to the Lewis and Clark Expedition of 1804–1806, the declaration of America's Blue Frontier can be traced to a 1972 Central Intelligence Agency cover story.

Of course, I had no way of knowing that on a dark summer night in June 1977, as I watched the *Deep Sea Miner Two*, a 560-foot, 20,000-ton, converted Liberian ore carrier slip out of San Diego harbor. The dock area around the Tenth Avenue Terminal had been sealed off to curious onlookers hours before the ship's unpublicized departure. I stood alone outside the barbed wire–topped hurricane fence, taking notes as a pair of uniformed guards on the pier watched me, speaking nervously into their walkie-talkies. The crew members who tipped me off to the ship's departure were warned not to discuss their destination or the nature of the tests to be conducted. For several of them who had served in the navy, it was not unlike going on a WESPAC (Western Pacific) deployment, except that this ship, with its strange radarlike white geodesic dome covering the derrick at midship and its pipe storage racks and air compressors crowding the topside and superstructure from bow to stern, looked far more peculiar than any of the warships tied up at the Thirty-second Street Naval Station. The ship, I discovered, was owned by Ocean Mining Associates of Gloucester Point,

Virginia, a joint venture of U.S. Steel, the Sun Company (formerly Sun Oil), and Union Minière of Belgium. The consortium was one of several formed in the early 1970s in response to news that Howard Hughes's Summa Corporation was committing hundreds of millions of dollars to an unprecedented deep-sea venture.

The *Deep Sea Miner* would be working 1000 to 1500 miles southwest of San Diego. There, in waters up to 20,000 feet deep, using dynamic positioning props to keep the ship steady, crewmen would sink dredges attached to several miles of pipeline. Compressed air would then be fed into each pipe to create a suction that would pull manganese nodules off the bottom of the sea. It was dangerous work, comparable to dropping a long, potentially explosive siphon from a blimp that was trying to remain stationary in gusting winds and clouds 4 miles above the earth. If the pipe string ruptured, the pressure could turn sections of broken pipe into steel spears, impaling ship and crew. Still, like many dangerous shipboard operations that take place every day on America's Blue Frontier, the *Deep Sea Miner* managed to get the job done without incident, recovering several tons of nodules.

A manganese nodule itself is fairly unimpressive. You can probably see one at a local oceanographic center or university marine lab. Up to 8 inches across, it lacks the speckled shine of a good mineralized throwing rock, looking more like a lump of soft coal. This may be because its growth pattern is less like that of the hardened spawn of the earth's upthrusting mantle than of an oyster pearl. It precipitates minerals out of the surrounding sea around a small nucleus such as a shell shard or shark's tooth. These accreted nodules are said to cobble parts of the deep ocean floor like the fabled streets of El Dorado and are invariably described in news reports as potato-sized, although you would have to be talking about those little gourmet numbers, not an Idaho spud. Their mineral content includes manganese, copper, nickel, and cobalt. And while most assays of nodules tend to be low, their total volume is truly staggering.

Samples of manganese nodules have been found and analyzed for their mineral content as far back as the cruise of the HMS *Challenger* in 1872–1876. Carrying out the first modern oceanographic expedition, *Challenger* used rope dredges to retrieve the slime-covered lumps from 5000-meter depths in the mid-Pacific.

By the 1960s the Scripps Institution of Oceanography and other research stations were involved in the Moho and Deep Sea Drilling Projects, attempts to drill through the ocean floor to the earth's mantle—big science projects conceived as a way to match the drama and attention space exploration was then receiving. Using converted oil-drilling ships with multi-screw dynamic positioning, they developed many of the techniques that are being used today for deep-sea oil exploration in the Gulf of Mexico. At the time these techniques were also seen as having potential for manganese mineral mining. On September 21, 1967, the Maltese ambassador to the United Nations, Dr. Arvid Pardo, gave a three-hour speech to the UN General Assembly, suggesting that the deep ocean floor, with its vast mineral wealth, be seen as "the common heritage of mankind," an idea adopted by the UN in 1970.

Then, in November 1972, Howard Hughes's newly formed Summa Corporation announced the launch of a massive 619-foot ship, the Hughes *Glomar Explorer*, to begin commercial recovery of deep-sea nodules. Dozens of enthusiastic articles on the project began appearing in major mainstream and scientific journals around the world. Few doubted that while the economic risks were enormous, the reclusive billionaire was just the type of man to make a daring investment in order to corner a new market and lead the entrepreneurial pack. What the deployment of the Hughes ship would actually lead to was not commercial mining of the seas but the first major remapping of the world's oceans since the sixteenth century, when Queen Elizabeth I of England gave up on church-negotiated ocean boundaries and endorsed the Dutch concept of mare liberum, or free seas.

For Russian and American submariners during the Cold War, the seas were anything but free; they were a cold, dark, claustrophobic place in which to hunt and hide, where submariners were blind as whales in black water but with far less sophisticated propulsion and sonar systems.

In March 1968 a diesel-powered, nuclear-armed Golf 11 Soviet submarine, the *Red Star*, sank in the North Pacific. A venting problem while recharging its storage batteries probably caused the powerful explosion that tore open its stern plates, sending seawater surging into the tail section of the craft, pulling it backward and downward in a deadly fast spiral. Within moments the *Red Star* reached crush depth before plowing into the soft bottom more than 3 miles down, traveling close to 200 miles per hour on impact. Her entire crew of some 100 men perished. They were among many submariners from both sides who would be crushed, drowned, burned to death, or irradiated during the Cold War, including, only two months later, 99 men aboard the USS *Scorpion*, lost in the depths of the Atlantic. While Moscow had no idea where or why its submarine disappeared, the U.S. Navy, using a combination of its top-secret $16 billion Sound Surveillance System (SOSUS) undersea hydrophones and intercepted satellite transmissions, was able to key in on the area where the Soviet sub went down.

In their definitive book on Cold War spy submarines, *Blind Man's Bluff*, reporters Sherry Sontag and Christopher Drew describe how the navy, in a major intelligence coup, had the spy submarine USS *Halibut* pin-point and photograph the sunken *Red Star* using cable-winched sonar arrays and still cameras (that also showed the skeleton of a sailor in a sheep-skin coat lying outside the wreck). The Russian sub contained code books and encryption gear and three nuclear missiles and their guidance systems, all valuable booty if they could be recovered. Leaders of the navy spy pro-gram suggested returning to the *Red Star* to do some silent salvaging, using C-4 explosives and a primitive version of today's ROVs (remote-operated vehicles) lowered by miles of cable from a stealthy American sub floating above. They were shocked when their plan was rejected and in its place the CIA proposed building a massive salvage ship to scoop the obsolete Soviet diesel boat whole from the ocean floor. The CIA plan (code-named Project Jennifer) went forward with hundreds of millions of dollars in contract pay-ments going to Hughes, Lockheed, and Global Marine, an oil rig platform manufacturer, to build the gargantuan spy ship *Glomar Explorer*. Eventually, more than 4000 people were employed in the construction and operation of the *Glomar*. Outraged navy men, including spy sub coordina-tor John Craven, quietly complained that the Nixon White House was using the CIA project to pay off its political patrons and California defense contractors before the 1972 elections. The fact that Howard Hughes was a long-time secret contributor to Richard Nixon's campaigns, as well as hav-ing provided a highly suspect loan to the president's erstwhile brother, Donald, added weight to the charge. The public unveiling of the giant spy ship cum ocean-mining vessel on election day 1972 may have even reflect-ed someone's cynical sense of humor.

Several possible cover stories for the *Glomar* were considered, including the idea that it was going on a treasure hunt, looking for lost Spanish galleons full of gold. But the CIA worried that such a romantic tale, linked to a mysterious figure like Howard Hughes, might attract too much atten-tion from congressional CIA oversight committees, among others. So, the cover story chosen was that of a manganese mining ship.

While Howard Hughes was working with the CIA, the Securities and Exchange Commission was investigating Hughes's financial empire and the takeover of Air West, a small regional airline. CIA overtures to have the SEC investigation sidetracked failed. The SEC was about to obtain a court order for Hughes's files when, on June 4, 1974, there was a mysterious break-in and robbery at his Los Angeles communications center. The files the SEC was after, along with $68,000 in cash and a memo discussing Hughes's work on Project Jennifer, disappeared. Having been denied access to the information it was after, the SEC investigation quickly ground to a halt.

At around the same time the *Glomar Explorer,* having completed a
series of sea trials, was launched on its "mining" mission.

Two weeks later about 5000 delegates from 48 nations met in Caracas,
Venezuela, for the United Nations Law of the Seas (LOS) Treaty convention. The talks, which had been going on in a desultory fashion since 1958,
went into overdrive as news of Hughes's half-billion-dollar ship heading
out to sea hit the assembly with the force of a tropical cyclone. Many delegates were outraged by the idea that the *Glomar* might soon begin sucking up thousands of tons of mineral-rich manganese nodules from the
ocean's depths. Those dark lumps were suddenly seen as a trillion-dollar
godsend by the UN's poorer nonaligned nations, organized as the Group
of 77 (G-77). Looking to the 1970 UN resolution declaring the seabed "the
common heritage of mankind," G-77 called for the establishment of a
UN-administered international seabed authority, labeled the
"Enterprise," to participate in the expected mining bonanza. The Soviet
Union, sensing a no-cost opportunity to align itself with the Third World
and against the United States, quickly endorsed the G-77 plan (it would
later abstain from a final agreement).

As the mining debate was heating up, the CIA salvage operation was
turning into a costly bungle. On July 4, 1974, the *Glomar* arrived on station
over the sight of the sunken sub and began lowering a giant eight-fingered
claw and steel netting attached to a three-mile tether of connected lengths
of hardened steel pipe. Assembling the tether as they went, the ship's crew,
made up largely of roughnecks from the offshore oil industry, dropped the
claw through the *Glomar*'s wet well. It took them several days before they
reached bottom and were able to snag the submarine. Lifting it at 6 feet a
minute, they managed to raise it more than a mile before the badly damaged
craft fell apart. In the end, they were able to recover only about 10 percent
of the submarine and the bodies of six sailors, which they buried at sea. No
missiles, code books, or encryption equipment were recovered.

Back at the LOS convention, the U.S. delegation was arguing that by
producing new mineral wealth from the ocean depths, private corporations
like Summa were, by definition, contributing to the common heritage of
mankind. It was an argument the majority of delegates vehemently rejected.
The CIA could only take grim satisfaction in watching its cover story blow
up into a major political cause célèbre. Hoping to return for a second salvage attempt (that never happened), the CIA continued to place mining
articles in major media around the world, while the U.S. media ran feature
stories on the *Glomar* in *Business Week,* the *New York Times,* and the *Wall
Street Journal.* To maintain the story's credibility, the Summa Corporation
also entered into a business agreement with American Smelting and

Refining to process their recovered ore. They even sent a small ship out into the Pacific to collect several thousand pounds of manganese nodules for publicity purposes.

Even after the Law of the Seas conference ended, the cover story continued to reach new, influential audiences. "The Hughes goal, it is reported, is to mine some 3 to 4 million tons of nodules annually," the Congressional Research Service reported to the U.S. Senate in December 1974, going on to project the future value of U.S. deep-sea manganese mining at $534 million by 1985. "We are confident that this document will serve as a valuable tool in our deliberations and policy-making," Senator Warren G. Magnuson, chairman of the Senate Commerce Committee, noted on receiving the report. Two years later Magnuson would lend his name to legislation establishing America's first 200-mile claim on its ocean frontier, a domestic fishery law aimed at excluding Soviet bloc factory trawlers.

With thousands of employees working on Project Jennifer and whispers in Washington and Los Angeles of recent failure and boondoggle (on top of Nixon's Watergate scandal and resignation), it was inevitable that the true story of the *Glomar*'s mission would come out. The first exposé, with the run-on headline "U.S. Reported After Russian Submarine / Sunken Ship Deal by CIA, Hughes Told," appeared in the February 7, 1975, edition of the *Los Angeles Times*. The *New York Times* investigative reporter Sy Hersh was also onto the story but had cut a deal with CIA Director William Colby to hold off on publication. Soon all bets were off, and there was widespread media coverage of the failed mission.

Oddly, despite the unraveling of the *Glomar* cover story, many of the world's leaders were now convinced that large-scale ocean mining was feasible.

A number of mining consortiums had been organized by energy and mineral companies out of fear that Hughes's Summa Corporation would get the jump on them. Membership included the Sun Company, British Petroleum, AMOCO, Mitsubishi, Noranda, Kennecott, Union Minière, and International Nickel.

By the end of the decade deep-sea mineral wealth, and who would benefit from it, was a major source of international tension, even though the practical costs of developing open-ocean dredging, transport technology, ports, and processing facilities (along with a decline in world metal prices) guaranteed continued land-based mining for the foreseeable future.

Michael Molitor, a former Law of the Seas official with both the United States and the UN, recalls a visit he made to the Canadian headquarters of International Nickel during this time. "They had this world map with small lights in Guatemala and Africa and all these places they had mining rights, including a single light out in the middle of the Pacific. And there was this dial to turn that showed the price of nickel rising, and as you'd twist that dial all these lights would come on showing when these mines would become profitable, but no matter how far you turned that dial that one light in the Pacific never went on."

Still, at the UN mining fever was running high.

"Deep-sea mining was one out of nine or ten issues on the table, and the only one that was not resolved to the satisfaction of the Americans," recalled Elliot Richardson, former U.S. attorney general and ambassador to the Law of the Seas convention. After standing up to Nixon during the Watergate crisis, the former Republican lawman, five years later, was named to head the U.S. delegation to Law of the Seas by Democratic President Jimmy Carter. Carter hoped Richardson would be able to untangle the diplomatic mess created by the mining issue.

By 1980 Richardson had worked out a compromise agreement that would see a mixed corporate UN mining regime established, but in 1981 a new delegation appointed by the Reagan administration nixed any "giveaway" of deep-sea nodules, insisting that the ocean's resources belonged by right to those with the capital and technology to claim them.

"The Law of the Seas [treaty] would have been signed off by the Carter administration if he'd been reelected, but the conservative ideological baggage brought in by Reagan on the mining issue forced a shift and those negotiations were stranded," Richardson told me.

University of Southern California Professor Robert Friedheim, a former navy consultant on LOS, agrees: "Richardson had negotiated everything the U.S. wanted but for ideological reasons the Reagan team rejected any reference to 'the common heritage of mankind.' That threw the process into a state of complete chaos."

Although the Law of the Seas convention eventually established 16 solid agreements on freedom of navigation, maritime commerce, maritime law enforcement, marine environmental protection, and marine scientific research, the lack of a seventeenth provision on mining left everyone nervous. Fearing that Western corporations would soon begin mining off their shores, members of G-77, like a school of panicked mullet, quickly moved to assert control over nearby marine fisheries and oil, gas, and minerals by declaring 200-nautical-mile (230 statutory miles) Exclusive Economic Zones.

Domestically, the debate over this radical EEZ concept had been going on at least since Chile, Ecuador, and Peru made 200-mile fisheries claims back in the early 1950s. The U.S. Navy, State Department, and distant-water tuna fishermen formed a freedom of the seas coalition opposing

EEZs. They saw them as a threat to global fishing and the navy's right of
passage through potential trouble spots like the Strait of Hormuz and narrow channels surrounding Indonesia. In favor of enclosure were the domestic fishing industry and the National Petroleum Council, which wanted to
move its offshore rigs into deeper water on the Outer Continental Shelf.
The Interior and Commerce Departments backed this group. After the
Law of the Seas negotiators recognized the right of free passage through
strategic straits for military ships, the navy abandoned its fishing allies, and
the balance of special interests shifted in favor of enclosure.

Although the United States refused to sign on to the final LOS Treaty
in December 1982, it wasted little time in following the example of other
coastal states, with President Reagan declaring the world's largest
Exclusive Economic Zone on March 10, 1983.

Two weeks later Secretary of Interior James Watt declared 70,000
square miles of this new ocean frontier open for mining. Wagons, ho! Watt,
who earlier attempted to declare a billion acres of the Outer Continental
Shelf available for oil leasing, was no friend of the marine environment, or
the dry one for that matter. A Christian fundamentalist, he favored unlimited development of America's natural resources. And while personal faith
should not be a factor in evaluating political appointees, Watt seemed to
take pleasure in aggressively arguing his beliefs as justification for what
were essentially nineteenth-century frontier policies. Asked in a congressional hearing why he was so determined to see public resources developed,
he responded that there was no point in long-term conservation because, "I
do not know how many future generations we can count on before the Lord
returns." Lanky, bald, and funereal in appearance, Watt was also not above
questioning the patriotism of those who disagreed with him.

After a run-in with the militantly moderate National Wildlife
Federation, he asked, "What is the real nature of the extreme environmentalists, who appear to be determined to accomplish their objectives at whatever cost to society? Is it to delay and deny energy development? Is it to
weaken America?" Among the gaffes that led to Watt's forced resignation
was his attempt to ban the Beach Boys, those wholesome muses of surf culture, from giving a free Fourth of July performance on the Washington
Mall.

The EEZ territory Watt wanted to open up to mining was the Gorda
Ridge, a recently identified site of deep-water volcanic chimneys, or "black
smokers," off the coast of northern California and Oregon. Like bumpy
free-standing tea spouts, these superheated chimneys spew dark concentra-
tions of percolating minerals, including zinc, iron, and copper, which form
towers of so-called polymetallic sulfides that can stand up to 150 feet tall

(researchers named the largest of these chimneys "Godzilla"). At the base of the towers can be found the same colonies of chemosynthetic clams, crabs, and strange feathery red-and-white tube worms that were first identified off the Galápagos back in 1977.

Reagan's hawkish Secretary of the Navy John Lehman argued that mining these deep seabed minerals would ensure secure domestic sources of strategic minerals for future defense. Environmentalists countered that the proposed lease sale was a vast giveaway of a public resource based on little or no scientific knowledge, with the added risk of destroying newly discovered life-forms. The mining industry, watching the price of land-based minerals heading south, was even less enthusiastic about the proposed sale than the enviros. Within a year the lease plan had been put on hold.

The Law of the Seas Treaty required ratification by 60 nations before it went into effect. This took more than 10 years. By then the United States and other Western nations had been talking with the UN about revisiting the mining provisions. This resulted in new terms more favorable to private development. On July 29, 1994, the Clinton administration, having achieved the deal that the Reagan administration demanded, joined what was by now 159 other nations in signing the Law of the Seas Treaty.

But North Carolina Senator Jesse Helms, chairman of the Senate Foreign Relations Committee, refused to hold hearings that would allow the Senate to ratify the treaty. As a result, in 1999 the United States, not being an official signatory, was thrown off the LOS commission assigned to establish the rules for ocean mining. The navy and State Department were also excluded from other LOS commissions on which they wanted to participate.

According to the senator's staff member responsible for Law of the Seas (who did not want to be quoted by name), the senator did not believe the Clinton agreement went far enough in protecting U.S. companies from paying mining royalties to "a bureaucracy where the U.S. doesn't have a greater vote than anyone else." I asked, "Would the treaty pass the Senate if Helms offered it up for a vote?" "His view is some other chairman can take it on as a priority, but it's not going to be his priority."

As Elliot Richardson told me a few months before his own death at age 79 on December 31, 1999, "Jesse Helms is doing significant damage to U.S. interests. But the navy and the Pentagon haven't been able to budge him, so I don't suppose anything will happen while he's there."

With the UN no longer claiming a share of "the common heritage of mankind," talk of strip-mining the deep seas once again came into fashion. By the year 2001 the island nation of Japan had a well-funded and ongoing commitment to developing deep-ocean technologies. At JAMESTEC

(Japan Marine Science and Technology Center) in Yokohama the government and leading corporations like Mitsubishi were leapfrogging the
unmanned *Kaiko* remote-operated vehicle and the *Shinkai 6500* (the deepest-
diving manned submarine in the world) down to the bottom of the sea. In
1995 the *Kaiko* came within a few feet of the U.S. Navy's 1960 record dive
site, almost 7 miles down in the Pacific's Marianas Trench (where the
robot's cameras recorded a sea slug and a shrimp). With few mineral
deposits of its own, Japan's interests seem linked, at least in part, to a desire
for resource independence through ocean mining. The United States,
Japan, France, Russia, and Germany have for some time been granting unilateral "exploration licenses" to mining companies in the international
waters of the mid-Pacific Clarion-Clipperton Fracture Zone, an area where
manganese nodules have a fairly high mineral content. The U.S. claim alone
in 2001 covered some 190,000 square miles.

In a reprise of the rhetoric of the 1970s, a 1995 issue of *Time* magazine
mentioned deep-ocean "oil and mineral wealth to rival Alaska's North Slope
and California's Gold Rush," while Karl Jugel, then head of the ocean
minerals division of the National Oceanographic and Atmospheric
Administration (NOAA), claimed the beginning of large-scale mining of the
seas was just a question of "when the nickel market is going to recover."

The movie *The Abyss* portrays a technically credible picture of a mid-
twenty-first-century mining operation, including a tracked and human
occupied deep-ocean dredger/ore processor linked by slurry and communications lines to a surface support and ore transfer ship.

Still, the future is rarely what the common wisdom projects it to be.
Today, many Americans feel that what is technologically feasible does not
always translate into what is healthy or desirable. The oil and gas industry,
for example, has given up new exploration off the East and West Coasts of
the United States, including Alaska's salmon-rich Bristol Bay, not because
there are not large reserves of oil there, but because popular pressure and
changing environmental policies require they look elsewhere. Similarly,
concern over preserving the ocean's health and genetic diversity based on
new scientific knowledge about the deep sea may scuttle the "inevitability"
of large-scale ocean mining.

Deep dredging or suction mining for manganese nodules even far out at
sea can be expected not only to cause heavy mortalities among slow-breeding
bottom dwellers but also to suffocate surrounding benthic communities,
according to a number of studies. Every 10,000 tons of nodules recovered
each day would generate 40,000 tons of sediment in the water column.
These sediment plumes, carried by the currents, could remain suspended
in the water anywhere from two weeks to 49 years (the time it takes for cer-

tain surface nutrients to settle on the deep bottom). The creation of this ocean smog could restrict light penetration and reduce planktonic growth that supports most life-forms in the ocean. It could also directly contaminate open-ocean fish stocks. Current mining claims overlap the migratory routes of yellowfin tuna and swordfish. Mining would also requires centrally located mining smelters and onshore processing facilities in areas like Hawaii.

Hawaiian waters themselves have been found to contain manganese crusts rich in cobalt. However, when the Department of the Interior's Mineral Management Service called a public hearing on leasing these manganese seabed crusts in 1992, close to 1000 angry citizens turned out on the Big Island, some dressed up as whales and dolphins. From native advocates of Hawaiian sovereignty to representatives of the tourist industry, fishermen, flower growers, surfers, and retirees, they spoke out forcefully against any type of mining proposal, arguing that the ocean already provided them with an economy and lifestyle that they were unwilling to risk for some new mineral bonanza.

Abraham Piianaia, a native scholar of Hawaiian history, suggested that a careful harvesting of fish and new medicines would be a more respectful way of approaching the deep sea. "My people are a maritime people," he explained. "They looked upon the sea not merely as a body of water, they looked at the sea as a source of food, a pathway to another place. The sea provided medicinal things that they used. The sea provided recreation when they needed that. . . . And now you read a lot of stuff, of Western science that says the source of all life is the sea. This may be true. And yet these simple people found that out years ago."

America's deep oceans, once perceived to be a biological desert whose main value might be found in mineral mining, waste dumping, and nuclear weapons testing, have more recently been identified as a key factor in ocean circulation, climate, and productivity (through nutrient upwellings), as well as a rich habitat for a wide and wondrous range of life. In terms of volume, 99 percent of the earth's livable habitat is in the oceans. But only now is human exploration of the middle and deep-ocean frontier below 660 feet (the planktonic zone) beginning to yield a sense of just how much of that life we have yet to know or understand.

An observation camera rests on the deep-sea bottom off Monterey, California, recording the movement of sea cucumbers, mollusks, and clams gathered around a cold water seep, a methane-rich chemical vent not unlike the scalding hot geothermal vents found elsewhere. A large seven-gill shark cruises by the camera. Thin spiderlike creatures float like mosquitoes around the pressure-sealed lights. What appears to be a rock will sit in front of the camera for months on end. Suddenly, over a period of days it will grow appendages and begin to crawl out of view. Ethereal translucent jellies hover on the periphery of the lights, flashing bioluminescent signals at them like alien code from a Steven Spielberg movie.

Scientists estimate that there are at least a million new species of life undiscovered in the deep oceans, not counting *Architeuthis*, the giant squid. Until recently much of our knowledge of this life came in the form of pressure-exploded fish and torn-up jellies lifted from the abyssal depths in trawler nets. ROV robot submarines now use specially designed slurp guns and pressure vessels to capture and preserve live specimens of deep-sea life for display at places like the world-famous Monterey Bay Aquarium.

"There are trends in oceanography just like in anything else. This area's now like the hot bar in a college town. For awhile the place to work for a young scientist was Woods Hole in Massachusetts, then Scripps in San Diego. Now it's Monterey, Monterey, Monterey," says Bill Duoros, who, as manager of the Monterey Bay National Marine Sanctuary, is not exactly an unbiased source. Still, his claim resonates with many people in marine science.

Looking out across the cool blue waters of Monterey Bay is like that first sweet shock of recognition when you fall in love. The heart of the world's second largest marine sanctuary (after Australia's Great Barrier Reef), its near-pristine waters teem with life, from mobs of barking seals to spouting gray whales to shoals of shearwaters scudding across its surface like squall lines. Just offshore, its submarine canyon contains sheer cliff faces more than 6000 feet deep that gradually fan out onto 10,000-foot-deep abyssal hills. But until the 1990s maps of the canyon were not available to the public or civilian scientists.

"Beginning in 1983 we had run an EEZ mapping effort using multibeam sonar ships," says Skip Theberge, a 28-year veteran of the NOAA Corps (the smallest of America's uniformed services, with some 280 members trained at the Merchant Marine Academy in Kings Point, New York). Theberge was in command of the NOAA research vessel *Mount Mitchell*, which mapped Monterey's deep canyon, a place that turned out to hide more than rockfish. "What we did was so accurate the navy submarine community felt threatened, so it was all classified." When the maps were declassified at the end of the Cold War, Monterey's deep waters began drawing scientists like chum draws gulls.

Gary Greene, a research scientist and former director of the Moss Landing Marine Lab (operated by the California State University system), has been on the scene since before it was a scene. He remembers his first submersible dive into Monterey's canyon back in 1970. He was on board the *Nekton,* a small steel sub, along with pilot Larry Headley. Headley sat upright in the conning tower as Gary, then a graduate student in geology, lay on his belly by the forward observation porthole, his legs sticking between Larry's feet.

"We'd detected a deep hole. It turned out to be the bottom behind where a massive landslide had filled in part of the canyon. We landed on the side of this steeply inclined mud and sediment dam about 1000 feet down, at the sub's maximum depth. Then, as we tried to cruise along our tail began to drag. We assumed it was tangled up in a cable or something. You could unscrew the *Nekton*'s tail section and float free, but between me having to crunch up in the bow and Larry wiggling around with the wrench, our movement must have set off another slide. I was looking out the porthole and everything suddenly went black from this big mud cloud, and we could see the pressure and depth gauges rising as we were pushed downslope beyond the sub's limit. We were trying to blow ballast [with compressed air from a pair of scuba tanks] but nothing was happening. The prop was making a loud clanking noise and then something fell out of the rudder. Then it started to get lighter outside and we were rising and I could see the mud cloud below us as we broke free."

I asked if they went back down again. "We dove again that day, but not in the same place." He pauses. "The strange thing is Larry got killed the following week in that sub. He was the observer and Rich Slater was the pilot. They had raised a Chris Craft that had sunk off Catalina, but then the line securing it broke and the boat sank back through the water column and hit their sub."

"It came tumbling down stern first and hit my porthole and cracked it and then, because of the pressure, the glass just imploded," recalls Rich Slater, now co-owner of *Delta,* a similar two-person submersible. "I got hit in the face and knocked out. The sub filled up with water and went back to the bottom, about 260 feet down. When it hit I woke up underwater and got the hatch open, and we both got out of there real fast. I was unconscious when they found me near a kelp bed on the surface, my face all bloody, my eardrums blown out. I was deaf for months. They had to put a bunch of stitches in me. My face is still scarred from the glass." I notice the scars but they do not seem too bad, mostly they blend with the crow's feet and character lines accumulated during a life fully lived. "Larry was a better swimmer than me, but he had these rubber boots on that must have filled with water. We think that's why he didn't make it back up," Slater explained.

"I can appreciate the risk involved but it's also important to get down there. The more we know about the ocean, the better off we'll be," says Don

Walsh. Speaking from his ranch 30 miles inland from Coos Bay, Oregon, Walsh is one of just two men who have been to the very bottom of the ocean, 35,800 feet down in the Marianas Trench. That was back in January 1960, more than 40 years ago. No person has been back since, although the $50 million ROV *Kaiko* came close in 1995, sending video images of that slug and shrimp back to the surface through its control cable. (Since remote control radio signals don't work well underwater, most robot subs have to be connected to the surface by a powered tether.)

Walsh got to the bottom on board a far simpler vessel, the bathysphere *Trieste,* a 50-foot-long, gasoline-filled flotation hull atop a 10-ton steel observation chamber built by the Swiss inventor Auguste Piccard, who sold *Trieste* to the U.S. Navy (his son Jacques piloted the craft on its record-setting dive). The navy later used *Trieste* to help locate the wreck of the nuclear submarine *Thresher,* which sank in the spring of 1963 with 129 men onboard. In October 1999, at age 68, Don Walsh went 8000 feet down in the Atlantic to look at hydrothermal vents, aboard a Russian MIR submersible. "That was the alpha and omega for me," he said. "First seeing earth recycling itself in the trenches back in 1960 and now seeing the earth creating itself at these chimneys."

Today you can see the *Trieste* at the Navy Museum at the Washington Navy Yards. Visiting this inner space craft at the museum's old converted armory building, I'm surprised how strangely fragile it seems, like a thin-skinned aluminum blimp above a heavy metal observation sphere with a tiny glass viewport. I accidentally pull a hose loose and quickly reattach it, glad we're on the surface. Compared to the *Spirit of St. Louis* or *Gemini* space capsule at the Smithsonian Air and Space Museum, which are seen by some 10 million visitors a year, this pioneering vessel is a neglected relic of one of America's boldest explorations. This is particularly odd when you consider that hundreds of Americans have followed Alan Shephard and John Glenn into space, but only one American has ever been to the lowest point on the earth's surface.

"There wasn't much of a navy selection process," Walsh chuckles. "I was one of only two volunteers. When I went down there it was one of the last geographic frontiers on the planet, but today we don't see much interest in making the national investment needed to explore the deep oceans."

A 1996 report from the National Research Council entitled *Undersea Vehicles and National Needs* agreed, reporting that support for ocean research (and the undersea vehicles needed for it) "has declined steadily in the United States, while the nation's need for knowledge about the oceans has grown." Little has changed in the years since.

Scientists are only now beginning to appreciate what the deep oceans might offer. Cold hydroseeps discovered in the Monterey canyon are being explored as sources of new bacteria and pharmaceuticals. The relation of these methane-rich seeps to earthquake faults is being studied by geologists from the United States and Japan, while fisheries experts are only beginning to understand the marine food web that we are rapidly depleting (midwater jellyfish, for example, appear to make up the bulk of biological mass in the bay). Pollution dispersion and the role of the oceans in the creation of weather, climate, and as a carbon sink in a warming greenhouse world are additional areas of study being pursued.

"The great thing about the canyon is that within an hour of leaving dock you can be in 3000 feet of water," says Chris Grech, until recently chief ROV pilot for the Monterey Bay Aquarium Research Institute (MBARI), which is one of 27 marine institutes now located along the bay's crescent shore. Established in 1987 by the late David Packard, the management wizard who with partner Bill Hewlett founded the giant Hewlett Packard (HP) computer corporation, MBARI has become the premier U.S. institute dedicated to creating new technologies for deep-water exploration. With $230 million of Packard Foundation money invested to date, its sprawling 140,000 square-foot oceanfront facility at Moss Landing, California, is an emerging force in oceanography.

"When our father set up the family foundation, he challenged us to contribute to the world in some new way," recalls Julie Packard, a tall, slender woman with penetrating brown eyes. Since both she and her sister, Nancy, were marine biologists who shared their father's love of nature, it did not take long for them to suggest converting an abandoned sardine cannery in Monterey into a world-class aquarium that would bring the experience of the bay to the public. Once the aquarium, with its multistory kelp forest exhibit, surge machines, and in-house sea otters, became self-supporting (with more than two million visitors a year), David Packard decided to establish MBARI as a separate science institute. "Dad was very excited about the bay's deep water. He wanted to do something major as an engineering challenge, the challenge of getting down there intrigued him," Julie recalls.

The choice of whether to put development money into multimillion dollar manned submersibles or into ROVs was the next issue he had to consider. It was a debate that in earlier years had divided the Scripps Institution of Oceanography (robots) from Woods Hole Oceanographic Institution (manned subs). While many scientists have tended to favor use of America's aging fleet of climb-aboard submersibles, industry (oil and gas, fiber-optic cable-laying, and marine rocketry) has opted for tethered robots that can be loaded onto ships of opportunity and, when necessary, sent on "suicide missions" among the wrecks, blowouts, and other disasters that periodically befall offshore operations. "It's really not an either-or choice. It's more like a toolbox from which you should pick the right tools for the job," says Don Walsh.

Still, with his experience in the development of HP robots for manufacturing and hazardous duty (inside nuclear containment vessels), David Packard decided to focus his resources on developing a new generation of ROVs. These would emerge from a collaboration of engineers, computer designers, and the marine scientists who would advise them on what they needed.

Chris Grech sits in the pilot seat of his control room on board MBARI's twin-hulled $22 million ROV mother ship, *Western Flyer*. A short, wiry, piebald man with mustache and goatee, he could easily pass for one of Sir Francis Drake's pirate captains (who sailed these same waters some 300 years ago). I watch as he scans a bank of monitors, a computer touch screen, and the slave controls that can adjust the movement of this 117-foot platform ship to those of an exploratory robot 12,000 feet below.

I walk down a few steps into a large metal-walled room in the heart of the vessel and over to a hulking blue, green, and yellow ROV with the MBARI seven-gill shark logo on its side. The ROV, named *Tiberon*, is 7 feet tall, 6 across, and 9 deep and is equipped with eyelike stereo video cameras, strobe lights, force feedback gripper claws (that allow the operator to feel resistance), and slurp guns (for sucking in octopuses, jellyfish, and other live samples). It is suspended between an open moon pool for easy deployment and recovery and a huge spool containing 12,000 pounds of control wire. The robot itself looks like a police tank sitting on top of a bear cage. This description does not please *Tiberon* project director Bill Kirkwood. "The prototype was a 928 Porsche," he admits ruefully.

MBARI owns a second deep-diving ROV, the *Ventana*, which operates off the Point Lobos (a smaller ship known as the "Point Puke" to those who have to take it out in rough weather). Sailing four days a week, it uses a microwave relay to provide a live video link from its robot's underwater cameras to scientists and aquarium visitors back on shore who get to share in a real-time exploration of the canyon.

In February 1991, Grech, working down the coast from Monterey, piloted the ROV *Ventana* to the newly discovered wreck of the USS *Macon*, a navy dirigible that crashed and sank in 1450 feet of water in 1935. Built in 1933, the *Macon* was the largest aircraft in history, a 785-foot rigid frame airship that also acted as a sky-based aircraft carrier for five Sparrow-Hawk biplanes. The fighter-spotter planes were stored in an internal hangar (much like the ROV *Tiberon* is stored inside the *Western Flyer*). They flew off and were recovered with an air hook hanging below the dirigible. The loss of the *Macon* to a violent gust of wind that drove it into the sea, killing two of its crewmen (81 others escaped by life raft), marked the end of the great age of American airships. Appropriately, its rediscovery marked a new era for underwater exploration by equally exotic craft.

"As we searched for the remains of the ship, we were microwaving live images to Dave Packard's house in Big Sur," Grech recalls with a grin. "He held a barbecue for his VIP friends and various scientists to show them what this new technology could do and it worked perfectly. They saw it as it was happening, the debris field, the outline of the *Macon* planes. It was really cool and very strange, plus it kept the boss happy."

Monterey's submarine canyon is now yielding a bounty of new treasures. "We're discovering about five new species of animals a year with the ROVs," Grech says. One of these species, the *Calyptogena packardana* clam is named after David Packard (who certainly contributed enough "clams"). "The public doesn't have a concept, not a clue about all the animals down there," the ROV pilot smiles, shaking his head in amazement.

But the biotech industry does. In Monterey and across the Blue Frontier the discovery of new life-forms has set off yet another wave of mining fever, this time involving bioprospecting for drugs, medicines, and microbes.

More than half of our present medicines derive from terrestrial plants and animals, including penicillin cultured from a common mold. But as society abuses antibiotics, feeding them not only to sick children but also to cattle, chicken, and farmed fish, more resistant strains of bacteria emerge, posing a troubling new threat to human health. In addition, traditional threats from viruses, cancers, and many other maladies continue to frustrate medical science. And so, as Abraham Piianaia suggested, we have begun to look back to the sea.

For years, ocean science centers like the Woods Hole Marine Biological Laboratory (MBL) in Massachusetts and the University of Miami's Rosenstiel School of Marine and Atmospheric Science have used marine animals for biomedical research. MBL discovered limulus amoebocyte lysate (LAL), a blood derivative of horseshoe crabs that is used to test for bacterial toxins. They also use horseshoe crabs to study vision (the helmet heads of the crabs have about 1000 large rods and cones per eye versus the 300 million you are using to read this). Toadfish, with well-developed inner ears, are used to study balance. Two toadfish even got to ride along with John Glenn on his final mission into space.

At the University of Miami I'm led on a tour through a stucco ware house full of sea hares (think guinea pig–sized, frilly, green sea slugs). Their big ganglia make them useful recruits for the study of brain and memory. I'm also shown pools of nurse sharks being raised for nonvoluntary hospital work (despite the media myth, sharks do get cancer, just not as often as smokers). In years past the CIA took an interest in venomous sea snakes and deadly stonefish. More recently, orthopedic surgeons discovered that porous hard corals make strong artificial bone implants that are not rejected by the body. The National Cancer Institute (NCI) Natural Products Branch recently began expanding its marine life collections. Since Taxol, a compound that freezes cancer cell growth, was first derived from the bark of yew trees in the early 1990s, seven other compounds with similar effects have been identified. Six of them come from marine organisms, including soft corals.

Other promising discoveries include anti-inflammation chemicals from sea feathers, virus-killing proteins from sea grass molds, and the bioremediation (toxic cleanup) potential of the bacteria *Beggiatoa*, which allow chemosynthetic clams and seaworms to convert hydrogen sulfide into energy.

The hardy microbes that thrive in the extreme conditions of heat and pressure around deep-sea vents and chimneys have become a favorite target of bioprospectors since the discovery and patenting of *Pyrococcus* (for DNA fingerprinting) in the late 1980s. While these microbes may hold great promise for drugs and industrial and agricultural processes, they are hard to collect and hard to work with once they are brought to the surface.

"Of all these microbes, you can only culture around 1 percent of them. But you can take a beaker of water or sediment containing these microbes and [using genetic engineering] do gross DNA extraction," explains Dr. David Newman, of the NCI Natural Products Branch. "We can also take gene clusters from one of these organisms and, like those plastic pop-it beads from the 1960s, reassemble them in a different sequence and see what we get."

With only 1 percent of federal research and development money for biotechnology being directed to the marine environment (where 80 percent of the planet's life-forms exist), a number of pharmaceutical companies have chosen to fund their own university-based ocean research. This has created situations rife with potential conflict of interest. Zeke Grader, executive director of the Pacific Coast Federation of Fishermen's Associations, served on the University of California's Institute of Marine Resources Advisory Committee from 1983 until 1989, when he quit.

"What finally got to me," he recalls, "I went to a meeting at Scripps, where the scientists were saying they weren't going to pursue this one promising find involving marine plants because they couldn't patent it, and the research was being funded by the pharmaceutical companies. And I said, 'What if it's a cure for cancer? You're not going to study it 'cause they can't make money off it? You're a public institution!' "

With vast potential for profit as well as progress, marine microbes and other life-forms retrieved from public waters using taxpayer-funded tools (like navy-owned academic ships and submersibles) might be expected to provide a direct return to the U.S. Treasury, perhaps through some kind of royalty payment such as exists for offshore oil and gas. But to date there has been no discussion of this in Congress. If Congress were to act, it likely would establish a token fee acceptable to the drug companies, given the huge sums industry spends to influence those kinds of decisions. In 1998, for example, pharmaceutical and health products companies spent more than $165 million lobbying Washington and an additional $13 million on that year's congressional elections.

Another little-examined issue is what the risks are when genetically altered species developed for aquaculture and other purposes are intro-duced into the marine environment like so many loose pop-it beads spilled on a dance floor. Researchers at the University of Alabama at Birmingham have cloned the gene for the blue crab's molt-inhibiting hormone (MIH). After blue crabs molt naturally (having outgrown their old shells), they briefly become softshelled and can be eaten whole. The researchers inserted the molt-inhibiting gene into insect cells in order to replicate large quanti-ties of MIH. They hope to use this material to find a way to genetically block the hormone's release. If they succeed, they could force crabs to molt on command, providing a year-round supply of softshell crabs for the seafood industry. The risk, according to critics of the biotech industry, is if such an MIH blocker gene got loose in the wild, it could turn blue crabs into soft targets for predators and disease, spelling disaster for the crabs, the ecosystems they inhabit, and the fishermen who depend on them. While public policy discussion of these issues lags, technology-driven science and exploration leap forward with the exuberance of wild dolphins in a ship's bow wave.

In 1998 the navy-owned University of Washington research vessel *Thomas G. Thompson* left Puget Sound heading for volcanic chimneys 200 miles off the coast. These formations are part of the Juan de Fuca Ridge, located northwest of the Gorda Ridge area, which was proposed for mining back in 1983. The University of Washington and the American Museum of Natural History in New York were cosponsors of this expedition. A video crew from the PBS television show *Nova* was also along to record the adventure. A year earlier the ROV *Jason* and Woods Hole's submersible *Alvin* were used to map and inspect several promising chimneys. Using transponders to help place the ROV, a series of sonar scans and stereo video

and digital pictures were taken and then turned into a 3-D mosaic map by
computer imaging software. The *Alvin* then went down to the seafloor to
collect tube worms and to allow scientists to take a firsthand look at the
chimney life, including giant spider crabs, white crabs, snails, sea stars,
skates, rattail fish, and assorted hangers-on. Biologists suspect that some of
these lowlifes could represent the first ecosystem on earth that goes back
almost four billion years.

On their return journey, the researchers carried the Canadian ROV
Ropos, which, like many a Canadian, knows how to handle a chainsaw, in
this case an underwater Stanley saw equipped with diamond-studded teeth.
One and a half miles down they snared and sawed off several chimney tops
(now on display at the American Museum of Natural History). They then
slowly reeled them to the surface using 8000 feet of yellow line. When the
third chimney broke apart coming on deck, they found its interior full of
exotic microbial life and shiny with yellow flecks of chalcopyrite, known to
an earlier generation of frontier prospectors as fool's gold.

While the promise of quick and easy riches from the oceans' depths has
not been realized, real treasure may yet be found as a result of the CIA's
Glomar Explorer fiasco of the 1970s. This treasure may prove greater than
military intelligence, copper, cobalt, or new microbes for the lab. For in
recklessly laying claim to an ocean wilderness larger and rougher than any
previous frontier in our nation's history, we have also stumbled onto an
opportunity to reclaim an important part of our past. The American oceans
remain, much like Patriot's Bridge at Concord, the battlefield at
Gettysburg, the Rocky Mountains, or the Golden Gate, an essential element in defining who we are as a people.

[Chapter Two] From Sea to Shining Sea

Sullen fires across the Atlantic glow to America's shore,
Piercing the souls of warlike men, who rise in silent night.

—WILLIAM BLAKE, AMERICA: A PROPHECY, 1793

Give me your tired, your poor,
Your huddled masses yearning to breathe free,
The wretched refuse of your teeming shore.
Send these, the homeless, tempest-tost to me

—EMMA LAZARUS, FROM THE NEW COLOSSUS,
INSCRIBED ON THE STATUE OF LIBERTY

My best memories of childhood growing up on New York's Long Island Sound involve water—standing, brackish, and salty. I remember the swamp in Douglaston, Queens where my friends and I used to play every day after school. We'd cut paths through the cattails and rushes that grew higher than our heads, startling rabbits and large pheasants that would burst into flight in front of us, setting our hearts beating to the flurry of their wings. Here we built reed forts and recreated the battles of Francis Marion, the Revolutionary War hero known as the Swamp Fox, using sticks and dirt grenades in place of musket and ball. In winter we would ignore adult warnings, sliding across the ice that formed where the swamp met the edging waters of the sound. One winter day two of my friends fell through the ice into freezing waist-deep water. After I helped them crawl out, all three of us hid in

31

the furnace room of one of their apartment house basements, drying clothes and bodies and taking a pledge of self-protective silence lest our parents find out.

Our fourth grade teacher, Mrs. Olson, was the daughter of a Long Island sea captain and used to bring mementos into class of America's maritime past, including a 6-foot serrated bill from a giant sawfish her father had caught. Mrs. Olsen tried to instill in us a sense of what it meant to live on an American island, even one rapidly turning into a vast network of metropolitan bedroom communities.

Our town's class divisions were also marked by water, by whether you were a member of the dock or the club. The country club sat on a hill a few blocks in from the shore and was open to the community once a year during Strawberry Festival. Although this was a good time for pie-eating contests, carnival games, and razzing girls, it also took much of the mystery away from the club. Having been given a chance to look over the tennis courts, mowed lawns, and chlorinated swimming pool, those of us who were members of the dock, knew we had the better of the deal.

The dock extended into Long Island Sound on creosoled wood pilings from a narrow, rocky beach below a seawall. At its end was a raised platform connected to three pontoon floats. Here we would practice cannonballs off the pilings or play dibble, diving after popsicle sticks that someone would release underwater or hide under a pontoon. Wading around the shallows was another kind of adventure, searching the muddy water with our feet for the primitive armored shapes of horseshoe crabs and then lifting them up by their spiky tails for closer inspection. Early on our boys' culture was divided between those of us who defended the rights of horseshoe crabs to be played with and skipped across the water and the older "hoods" who liked to imprison them in rock corrals and then smash their shells with heavy stones. I remember after one fight in which by dint of numbers we vanquished a group of "hoods," a gray-haired eel fisherman came over to congratulate us, explaining how sometimes you have to fight for creatures who can't fight for themselves.

I also remember Hurricane Donna, which swept up the eastern seaboard when I was nine. My mother drove me and my sister down to a street overlooking the dock. We parked behind the town's two police cruisers, with a hundred other townspeople in the whipping rain, watching the gray sound's white-capped waves smashing against the seawall, watching the dock swaying side to side, until the slashing winds tore the wooden slats and rails off their pylons, hurling them into the sky like a runaway fenceline. I thought it was the coolest thing I'd ever seen.

By the time I entered junior high school, the swamp had been filled in and covered over by asphalt and red brick tract homes, and the newly contaminated waters of the sound were closed to fishing and swimming. My memories of these natural pleasures lost are not unique but rather the common currency of people raised along America's shoreline during the second half of the twentieth century, a period of rapid transition on our Blue Frontier, a period during which our sense of connection to the oceans around us eroded like a walled-off beach turned to rock cobble. It wasn't always that way, nor need it be again.

From its first human settlements, North America has been a maritime culture. Fishing was key to a number of early arrivals such as the Nez Percé, Tlingit, and Chinook of the Pacific Northwest who lived off salmon and believed humans came to the land on the back of a killer whale. The Eskimos hunted in sealskin kayaks, while the Makah used 12-man canoes to harpoon gray whales that migrated along the west coast more than a thousand years before European colonists began whaling off the island of Nantucket. In what would become California, the Miwok, Yurok, Chumash, and other tribes traveled the coastline in wooden plank boats gathering mussels and abalone along the marine terraces, warmed themselves in sea otter fur, and hunted giant 16-foot sturgeon in the shallows of San Francisco Bay. Far to the east, the Penobscot, Pequot, and Iroquois fished the Atlantic's bays, islands, and estuaries, using wooden fishtraps, spears, and hooks baited with lobster. The Narragansett of New England were the first to fashion wampum beads for trading from the dark, mother-of-pearl insides of quahog clamshells. Far to their south, the Seminole and Calusa of Florida depended on the Everglades "Pa-hay-okee" (grassy water), as well as the clear, sparkling, low-nutrient waters of Florida Bay for much of their sustenance, using nets, traps, and hooks to catch sheepshead, sea trout, and drum.

Early European explorers and settlers were awestruck by the wealth of fish, shellfish, seabirds, and marine mammals found along the eastern seaboard. When the English navigator Henry Hudson arrived in a great wilderness harbor in 1609 he found local Indians fishing for shad off a small sandy island they called Kiosk or Gull Island. The Dutch and English later called it Oyster Island, in tribute to its rich shellfish reefs, while still later it became known as Ellis Island, the port of entry for more than 12 million of America's oceanic immigrants.

To the north, the Pilgrims were surprised to find Cape Cod and Plymouth harbor alive with whales when they arrived aboard the *Mayflower* in 1620. By then, fleets of European Basque fishing boats had already been hauling cod off the nearby Saint Georges Bank for more than a century. In 1649 the governor of New Amsterdam wrote home to Holland, telling of 6-foot lobsters in the local waters around Manhattan, while an Englishman spoke of catching giant cod simply by lowering baskets weighed down with stones and hauling them back up. In 1699 Pierre Le Moyne, Sieur d'Iberville, his brother, and 13 others landed at what would become Biloxi, Mississippi, on the Gulf Coast, found an abundance of game "and some rather good oysters," built a fort, and established the French territory of Louisiana, still renowned for its seafood.

From New England to the great bay the Indians called "Chesepiooc" (Chesapeake), whose oyster reefs posed a hazard to navigation, and on south to the parrot- and turtle-rich Georgia isles, the settlers were overwhelmed by the blessing and abundance of the seashore, which helped sustain them through rough, hardscrabble times.

The English colonial authorities were also much taken with New Hampshire's tall and majestic white pines, which made superior stock for the Royal Navy's masts and spurs. Under the White Pine Acts of 1722 and 1729, the British crown established the "broad arrow policy," marking the best trees (those averaging up to 40 yards by 40 inches in diameter) with three blazes, creating for the navy the first public land reserves in America. But lumbermen and mill owners among the colonists saw little profit in letting the crown export raw logs to England to become ships' masts (also known as "skyscrapers" before buildings grew taller than ships). They cut 500 marked trees for every one that got shipped to the mother country. After the British surveyor General Daniel Dunbar tore down a mill full of blazed logs in the town of Exeter, New Hampshire, in April 1734, he and his men were set upon by a mob and driven away with small-arms fire in a pattern of resistance that would repeat itself with increasing ferocity over the next 40 years. When the lieutenant governor appealed to the state legislature to punish the culprits, he was stonewalled by taciturn legislators, many of whom made their living in the local timber trade.

Much of this timber went to shipyards then being established along the eastern seaboard. The colonial shipbuilding industry not only helped stimulate local fishing and commerce but also spurred subsidiary industries that sprang up around the shipyards, such as tar, rope, barrel-making, and ironworks (producing anchors, chains, and nails). Seaports like Boston, Newport, New York, Philadelphia, and Charleston became cultural and commercial centers for the colonists as well as centers of disaffection with English trade and conscription policies.

The colonists, although many were dedicated sailors and watermen, put up fierce, often violent resistance when it came to impressment, the Royal Navy's tradition of sweeping towns for able-bodied seamen. British captains were "mobbed," boats burned, and ships fired upon. Commodore Charles Knowles's efforts to sweep Boston Harbor for able-bodied men in

November 1747 led to two days of heavy rioting. In 1764 there were riots and gunfire in Newport, Rhode Island, after a deserter escaped (the British ship *St. John* was also shelled by the colony's gunners at Fort George). That same year four fishermen seized off Long Island were released after the British ship's master was taken hostage ashore, and there was additional rioting in Norfolk in 1767 and in Boston in 1768.

This physical resistance gradually mixed with talk of freedom, natural law, and unjust taxes, as yeoman farmers, sailors, and fishermen gathered in their homes and around their hearths, ofttimes built and heated with broad arrow pine. In their discussions and debates over the meaning of freedom they formed a new American identity. Not surprisingly, one of the first martyrs to the American cause was a former slave and able-bodied seaman, Crispus Attucks, who was shot down in the Boston Massacre of 1770.

In 1775 Edmund Burke warned his increasingly frustrated and bellicose colleagues in the British Parliament not to underestimate the strength and spirit of these troublesome Americans. "Look at the manner in which the people of New England have of late carried on the whale fishery," he suggested. "Whilst we follow them among the tumbling mountains of ice, and behold them penetrating into the deepest frozen recesses of Hudson's Bay and Davis's Straits, whilst we are looking for them beneath the Arctic circle, we hear that they have pierced into the opposite region of polar cold, that they are at the antipodes, and engaged under the frozen serpent of the South. . . . Nor is the equinoctial heat more discouraging to them, than the accumulated winter of both the poles."

With the outbreak of the American Revolution in 1775 came the creation of a coastal and Great Lakes navy. This force, with the exception of John Paul Jones, who favored leading his crews into direct battle with the Royal Navy, spent most of the war practicing a legalized form of piracy known as privateering. American schooners, with letters of permit from Congress, attacked British merchant ships and seized their cargoes (some 600 in all), while striving mightily to avoid one-on-one confrontations with well-armed British warships. This type of maritime guerrilla warfare, while risky, never lacked for volunteers, as most of the American fishing and whaling fleets were bottled up in their home ports by British naval blockades.

With U.S. independence in 1783, the fleets were freed up for global commerce, including the ongoing trade in human chattel, the infamous triangular slave trade between Africa, the Caribbean, and North America, which would continue until 1808, when it was officially banned. This trade in human misery had by that time provided much profit for New England's burgeoning shipping industry.

In the wake of the Revolution, Massachusetts cod fishermen were guaranteed access to Canadian waters as part of the Anglo-American peace accord. What was not guaranteed was the right to a defensible marine boundary for the new nation. In 1793 Thomas Jefferson, as secretary of state, announced a 3-mile territorial limit for the United States (the standard range of a cannonball at the time, or so Jefferson claimed). Still, Britannia continued to rule the waves and to impress U.S. sailors from American merchant vessels. It was this practice that led to the War of 1812, the British torching of Washington, D.C., and the 1814 siege of Fort McHenry in Baltimore Harbor, which inspired Francis Scott Key to write a resistance poem by the dawn's early light, which later became America's national anthem.

During this period fishermen, sealers, and whalers from towns like Nantucket; Gloucester, Massachusetts; and Stonington, Connecticut, began to gain fame as among the finest in the world (having the world's richest offshore fishery in Georges Bank didn't hurt).

Before the discovery of petroleum, whale oil was used to light the world's better homes and streets and was the primary lubricant of the machine age. New England whalers became the leading hunters of "Leviathan" (as recounted in Herman Melville's classic American novel *Moby Dick, or The Whale*). Whaling was a mainstay of the U.S. economy through much of the nineteenth century, and its impact was comparable to that of the post–Civil War railroads. By 1857 the town of New Bedford, Massachusetts, employed some 10,000 men working on 329 whaleships. Having emptied out the near-shore populations of whales early on, New England's whalers began hunting along the coasts of Africa and Brazil (today these same waters are being prospected by offshore oil companies).

Fur hunters were among the first to round South America's Cape Horn and join Russian trappers operating along the Pacific coast of North America, killing sea otters in vast numbers and trading their skins in China for tea, spices, porcelain, and silk. In 1820 Nathaniel B. Palmer, a 21-year-old Connecticut sealer commanding the small sloop *Hero,* became the first person to spot the Antarctic continent, while hunting the southern ice for fur seals. By the end of the 1820s Palmer and other hunters had killed some three million fur seals, selling their skins in China as fake sea otter. The sealers soon overhunted themselves out of business but in the process opened up Antarctica's Southern Ocean to commercial whaling.

Yankee whalers also began heading west around Cape Horn and out into the Pacific. The Hawaiian Islands, under the rule of King Kamehameha, became a major provisioning center for whalers, who would stop there twice a year, once in the spring before heading north to spend the summer hunting on the Sea of Japan and again in the fall before heading south to cruise along the equator. By midcentury some 700 whaling ships were operating in the Pacific. Hawaii's tropical Polynesian culture with its less rigid sexual mores would prove to be both a revelation and great temptation for many sailors from cold, Puritan New England (who in turn would prove to be a source of syphilis, cholera, and other infectious diseases for the native Hawaiians). When Yankee sailors jumped ship in the islands, they were quickly replaced by young Hawaiians, lifelong watermen and trained artisans whose finely etched whale-tooth scrimshaw carvings can still be found on display in maritime museums throughout the world.

Whaling was by its nature a boring, dangerous, and challenging profession in which small whaleboats from a mother ship were sent out on often rough seas to pursue and harpoon pods of whales. Harpooned whales often dragged the boats for miles in dangerous "Nantucket sleigh rides." Everything was expendable in pursuit of the whales—the harpoons, the rope, the line tubs, the whaleboats, even the whalers themselves.

Also onboard the whaling ships of the 1800s were large black iron kettles called try-pots, which were placed on a brick shelf or pile of sand and were used to "try," or render, whale blubber into oil while the ships were underway. In the western Pacific try-pots were recovered from storm-wrecked whaling vessels by cannibalistic Fijians who used them to cook their victims. This later became the basis for the cartoon stereotype of the cannibal cooking a missionary in a big kettle.

Whaling competition with Britain, France, and Russia eventually led to increased U.S. settlement of the Hawaiian Islands and the overthrow of the Hawaiian monarchy in 1893, but by then whaling was an industry in decline. This decline can be traced to several factors, including the first drilling for petroleum "rock oil" in Titusville, Pennsylvania, in 1859 (kerosene, distilled from petroleum, proved far superior to whale oil for lighting lamps); the outbreak of the Civil War in 1861 (Confederate warships sank 50 northern whaling vessels); and the Arctic ice disasters of 1871 and 1876 (in which 45 New Bedford ships and hundreds of men were lost).

Along with whaling, other maritime fortunes were made and lost in trade and transportation, especially after the Mexican-American War of 1846–1848 and the discovery in 1848 of gold at Sutter's Mill, California, which opened up the West Coast to rapid Anglo settlement. While clipper ships spent months racing around Cape Horn during the gold rush, Cornelius Vanderbilt came up with a quicker method of transporting miners to California, establishing his Accessory Transit Company of fast clippers, stagecoaches, and riverboats cutting across the Central American isthmus at Nicaragua.

Safety at sea also took a qualitative leap forward in the 1840s, when navy officer Matthew Fontaine Maury published the first reliable U.S. wind and current navigation charts, based on the logs of whalers and other American seafarers. As hydrographer of the navy from 1842 to 1861, when he defected to the Confederacy, Maury became known as "Pathfinder of the Seas." Equally impressive was the work of the civilian U.S. Coast Survey, which mapped the coastlines under the able leadership of Alexander Dallas Bache.

One of Bache's surveyors wrote to him from the Florida Keys: "You ask whether the growth of coral reefs can be prevented, or the results remedied, which are so unfavorable to the safety of navigation. . . . I do not see the possibility of limiting in any way the extraordinary increase of corals, beyond the bounds which nature itself has assigned to their growth."

The 1840s also marked the first great wave of postcolonial immigration, made up mainly of Irish fleeing the potato famine, along with Germans and Italians fleeing political turmoil that culminated in the European revolutions of 1848. Many of these immigrants arrived in steerage, named after the low-cost accommodations below deck by the stern rudder that steered the ship. Chinese also began arriving on the West Coast in response to news of the California gold rush. More arrived in the 1860s and 1870s to provide day labor as "coolies" (derived from the Chinese words *Koo* and *Lee,* which translate as "rent-strength"). They would help build the western link of the transcontinental railroad, while Irish immigrants worked the eastern section.

This early Asian immigration reflected the fact that China was a closer source of trade and cheap labor for California businessmen than the U.S. East Coast, if you had to sail around the Horn. The railroads solved part of this transportation problem by establishing a national infrastructure for the movement of people, beef, and buffalo skins. Still, California continued to ship its grain by sail around Cape Horn to Europe until the end of the century. In 1881 no fewer than 559 ships assembled in San Francisco Bay to load grain bound for northern Europe. Along with these commercial traders, the U.S. Navy was also feeling hamstrung by its inability to quickly move ships from one coast to the other.

By the 1850s naval strategists and engineers in both the United States and Europe were beginning to discuss the possibility of a transoceanic canal, either in Nicaragua or through the Darién jungle on the Isthmus of Panama, which was then part of New Granada (Colombia). U.S. interest in the canal was delayed, however, by growing domestic turmoil and increasingly violent conflict as the nation was divided between free and slave states.

The American Civil War, along with its fields of slaughter, saw an industrial and technological expansion that included large-scale steel production, the growth of railroads, and the introduction of steam power, gun turrets, and metal armor on American ships of war. This in turn led to the famous, if inconclusive March 9, 1862, "brown water" duel between the USS *Monitor* and the CSS *Virginia* in the shallows off Hampton Roads, Virginia. This deafening battle of ironclads marked the beginning of the end of wooden, wind-dependent warships and the rise of the heavy metal, engine-driven battleship fleets of the twentieth century. But, unlike the railroads, which expanded into the arid western frontier immediately following the Civil War, it would be another 30 years before the navy's leadership fully appreciated the new potential for changing the nature of warfare on the Blue Frontier.

Aside from western expansion, the post–Civil War era also saw the railroad companies opening up the seashore to vast numbers of "day trippers" and urban "expeditioners." Until then coastal holiday retreats were largely the provenance of wealthy society types and robber barons, of Astors and Vanderbilts who built their massive summer "cottages" in Newport, Rhode Island, or at similar enclaves such as Bar Harbor, Maine, and Avalon, New Jersey. Grand and elegant beachfront resort hotels like the Victorian Royal Poinciana in Palm Beach, Florida, and the many gabled Hotel Del Coronado in San Diego, famed for the giant black sea bass that could be caught by surf casting from the hotel's strand, also drew the cream of society.

At the same time, America's urban working classes and nascent middle class discovered that for the price of a train or trolley ticket they, too, could escape the broiling cities, if only for a day, and make their way to New York's Coney Island or Atlantic City and Cape May on the 127-mile-long New Jersey shoreline. Coastal areas like Cape Cod, Massachusetts, long dependent on fishing and shipping for their livelihoods, suddenly began to realize new economic potential in ocean-oriented tourism. Joseph Story Fay, the businessman who gentrified the fish rendering town of Woods Hole on Cape Cod, and Spencer Fullerton Baird, who established the first fisheries laboratory there, both arrived as railroad tourists in the 1870s.

On the Coney Island boardwalk, amusement park attractions with exotic sounding names like Steeplechase, Luna, and Dreamland would soon enter the American lexicon. In Atlantic City saltwater taffy became a popular treat identified with the boardwalk and the shore, while across the country by San Francisco's icy shore the fantastic Cliff House hotel and glass-enclosed swimming pools of the Sutro Baths offered thousands an opportunity to enjoy the beach without actually having to enter the cold and treacherous Pacific.

Working with the post–Civil War railroad trusts, the Cleveland rock-oil refiner John D. Rockefeller, along with his close friend and colleague Henry Flagler, were able to create the Standard Oil monopoly, using railroad shipping rebates to undercut their competitors. Soon oil production moved from Pennsylvania to east Texas and California, where the first ocean drilling took place off wooden piers in Summerland, a spiritual cult colony, just south of Santa Barbara. In Louisiana the newly formed Gulf (of Mexico) Oil Company began drilling in lakes and swampy coastal areas. As Rockefeller became more controversial, attacked by the press and investigated by Congress, he also became more philanthropic, establishing among other endowments the Rockefeller Foundation, which became a major funder of U.S. oceanographic research (which in turn would create and field-test many of the deep-drilling technologies used in modern offshore oil exploration).

In 1884, as Congress started funding construction of new, heavily armed cruisers, dreadnoughts, and battleships, the Naval War College was established in tony Newport, Rhode Island. Among its first faculty was Captain Alfred Thayer Mahan, a student of the Royal Navy and its links to empire. Mahan rejected the U.S. naval tradition of commerce raiding and brown water coastal defense as less than worthy of a great maritime power that, he argued, should be able to engage foreign fleets in direct battle. His book *The Influence of Sea Power upon History* (1890), electrified not only naval strategists but the public, including a young politician and future president Theodore Roosevelt, who became a lifelong advocate for Mahan and his theories.

The growth of a powerful blue water navy became not only a tool of late nineteenth-century empire-building but also an excuse for it. Coal-fired battleships needed far-flung deep-water harbors and refueling stations, went the reasoning. Also, the cost of maintaining a two-ocean navy could be greatly reduced by constructing a canal across Central America to facilitate rapid movement of the fleet.

In 1889 the idea of a coaling station and navy harbor in Samoa almost led the United States into a war with Germany and Britain (until a typhoon sorted out the vessels of the three nations, leaving only the British warship unscathed). In early 1892 the United States came close to war with Chile over a barroom brawl in Valparaíso that left two U.S. sailors dead. Then, in 1893, with the support of the U.S. ambassador, American planters in Hawaii overthrew Queen Liliuokalani. U.S. marines off the heavy cruiser *Boston*, docked in Honolulu harbor, were dispatched to the Royal Palace to protect the coup plotters. Pearl Harbor, Hawaii, subsequently became the major U.S. naval base in the Pacific.

In 1894 a U.S. warship exchanged fire with Brazilian rebels blockading the harbor in Rio de Janeiro. In 1895 the United States threatened war with Britain over raids from British Guyana into Venezuela, which the United States saw as a violation of the Monroe Doctrine of U.S. dominance in Latin America.

Finally, in 1898 the Spanish went too far. A year earlier, while the Spanish colonial army in Cuba was suppressing Cuban nationalist rebels, the U.S. Naval War Board began planning a war with Spain, including an attack on the Spanish fleet in the Philippines. In January 1898 the navy ordered the USS *Maine* to sail from Key West, Florida, to Havana, Cuba, on a "courtesy call." The wife of the *Maine*'s captain suggested, "You might as well send a lighted candle on a visit to an open cask of gunpowder!" On February 15, 1898, the *Maine* exploded in Havana Harbor, resulting in the deaths of 253 men. Many of their bodies were returned home aboard her escort vessel, the USS *City of Washington*. Whether the sinking of the *Maine* was caused by a mine or accidental detonation has never been determined, but at the time the Spanish were blamed and America went to war.

After several decisive naval victories, including the Battle of Manila Bay in the Philippines, the United States found itself in possession of an island empire incorporating Cuba, Puerto Rico, and the Philippines, where rebel insurgents, who had been fighting the Spanish for independence, were soon embroiled in a long and vicious guerrilla war against their American "liberators."

The end of the nineteenth century and the first two decades of the twentieth marked the second great wave of immigration, with the arrival of some 25 million of a total 65 million immigrants who came to America by ship. Almost 90 percent of today's Americans are descended from these shipborne immigrants, while another 10 percent were born elsewhere, part of a third great wave of Latin and Asian immigration that began in the 1970s. (Another 1 percent of Americans came much, much earlier by way of the Bering Sea land bridge.)

Unlike the wave of the 1840s, the turn-of-the-century immigrants were mostly from southern and central Europe. Like their predecessors, however, they were fleeing poverty and religious and political persecution. They included Italians, Greeks, Austro-Hungarians, Russians, Poles, and Ukrainian Jews like my father, Max Helvarg. A not untypical immigrant, 11-year-old Max arrived on Ellis Island on September 22, 1922, with his eight-year-old sister, Sue, and widowed mother, Liza. They were refugees from the civil war that followed the Bolshevik Revolution. They had hidden in an attic with 20 other people for a week while Cossacks raped and killed the other Jews of their village. After the attack they fled toward Romania, hidden under the straw of a hay wagon. Max, nine at the time, and Sue, six, got separated from their mother at the Dniester River. Max and Sue were later smuggled across the frozen river ice. They had to climb a steep cliff on the other side and walk across the cold stubble of fall cut wheat fields. Max carried Sue on his back. The spiky stubble tore his shoes and shredded the soles of his feet, so that he was leaving bloody footprints in the frost by the time they arrived in the village of Belz, where their mother was waiting for them. They waited there two more years before traveling on to Bucharest and the port of Constantsa, where they sailed for the United States on the liner *King Alexander*. It took 30 days to make the passage through the Dardenelles to Constantinople (now Istanbul), past Athens and Gibraltar, and across the deep Atlantic to New York City.

Arriving in the harbor, Max searched for the Statue of Liberty, but all he could see in the fog was the electric Wrigley's chewing gum sign on the Jersey shore. The first black person the Helvargs ever met was the immigration inspector on Ellis Island assigned to question them. He smiled and spoke to them in fluent Yiddish. Max and Sue stared at him in amazement. When Liza's brother arrived, he took them on the subway to his home in Coney Island. Max and Sue were certain they would never get out of that hole in the ground. Then they were out of it and riding in the light as the elevated train took them above the buildings and the river, the bright seashore, and the promise of freedom.

The year of the century, 1900, was marked by patriotic celebrations of America's new maritime empire. Rudyard Kipling, the British poet laureate of imperialism, smitten by the recent American invasion of the Philippines, encouraged the United States to "take up the white man's burden." Subsequent U.S. Marine landings in Mexico, Panama, Nicaragua, Haiti, and the Dominican Republic suggested that some in political power found this idea appealing.

Along with jingoistic celebrations, 1900 also saw the costliest natural disaster, in terms of lives lost, in U.S. history. That September the wealthy Victorian port town of Galveston, Texas, built on a sandy barrier island, was devastated by a storm-driven sea surge that killed more than 6000 of its citizens, about a third of the town's total population. In the wake of the disaster the citizens determined to stay and rebuild in harm's way. The Army Corps of Engineers, which had its origins in the construction of coastal forts and gun batteries, was enlisted to reinforce a 10-mile-long seawall erected to protect the town.

In 1903 and more than 1000 miles to the south, the United States helped organize a secessionist rebellion in Panama. The U.S. gunboat *Nashville* and a battalion of marines kept the Colombian army at bay as the insurgents consolidated power. Within a few weeks the United States had recognized the "revolutionary" new government and negotiated a treaty that granted the United States sovereignty over a 10-mile-wide strip across the length of the isthmus. French engineers had already spent years, fortunes, and thousands of lives in a failed attempt to excavate a transoceanic canal there, but 10 years after beginning their work in 1904, American engineers had completed the canal. This was in no small part thanks to the work of Dr. Walter Reed in Cuba and Colonel William Gorgas in Panama, who identified malaria and yellow fever as mosquito-borne infections and fought to eradicate these diseases among the canal's workers by eradicating the bugs and their breeding grounds. Still, more than 5600 workers died building the canal, mostly West Indians. If the French construction effort is included, the death toll exceeds 25,000. The completion of the Panama Canal was widely celebrated in the United States, with Pan-American fairs and expositions in San Francisco, San Diego, and other coastal towns, even as American troops began shipping out for the Great War in Europe.

Shortly before completion of the Panama Canal, Americans' faith in the new technological wonders of their age was briefly shaken by the April 15, 1912, sinking of the White Star liner *Titanic,* owned by American industrialist J. P. Morgan. Carr Van Anda, managing editor of the *New York Times,* reflecting on the sudden cutoff of the *Titanic*'s radio signal after its collision with an iceberg, concluded the ship had sunk and put that out in the next day's edition. His educated guesswork established the *Times* as America's newspaper of record.

As survivors reached New York on April 18, more than 40,000 people jammed the city's docks to witness their arrival. Swarmed by press and newsreel photographers, the survivors faces were splashed across newspapers and shown as flickers—movies—on screens across the nation. With the *Titanic* disaster the emerging media of still and motion picture news photography came of age, generating the first instance of a media frenzy. A more productive result of the disaster was the expansion of maritime safety regulations that mandated enough lifeboat space for every passenger, regular evacuation drills, better lookout systems, and 24-hour radio watches.

The *Titanic* tragedy was subsumed in the public consciousness by the more immediate horrors of World War I. The introduction of submarine warfare, and particularly the 1915 German sinking of the passenger ship *Lusitania* (which, along with a number of American passengers, was carrying contraband ammunition to Great Britain) again changed the nature of naval warfare. The U-boat foreshadowed more than 80 years of expanding submarine and antisubmarine warfare. The only comparable "revolution" in modern sea power came during the next war, when the Japanese attack on

Pearl Harbor and the Battle of Midway established the dominance of war planes and aircraft carriers over battleships and heavy cruisers.

The 1920s, along with a disarmament treaty that temporarily reduced the size of the world's navies, was marked by economic growth and prosperity across the United States. Mass-production technologies and the coming of age of the automotive and advertising industries saw America transformed into a consumer culture. Radios, wristwatches, electric washers, refrigerators, and processed foods sold through supermarket chain stores like A&P and Piggly Wiggly helped create a national middle-class identity. By 1922 Clarence Birdseye was quick-freezing fish fillets near the Fulton Fish Market in New York, while the sale of cooked-in-the-can sardines skyrocketed. The popularity of this original "fast food," along with canned tuna, shrimp, oysters, or, if you were poor, inexpensive canned salmon from Alaska, kept fishing docks and processing plants busy from Juneau, Alaska, to Monterey, California, to Lubec, Maine. The introduction of gasoline and diesel engines on fishing boats also allowed for the use of larger, heavy-bottom drag nets called otter trawls that created new fisheries for flounder, halibut, and other bottom-feeding flatfish. Catches increased rapidly, as did the number of fishermen. This fishing pressure, combined with environmental impacts from dams, water diversions, and other onshore activities, led to the collapse of a number of species. Some of them, like the Atlantic salmon and Chesapeake oyster, have yet to recover.

Separated from the U.S. Navy in 1919 and placed within the Treasury Department, the U.S. Coast Guard spent much of the Roaring Twenties chasing after Prohibition rumrunners, just as it spent much of the 1990s diverting patrol boats, cutters, helicopters, and falcon jets for the war on drugs. The year 1920 also saw passage of the Jones Act, which provided government subsidies to U.S. merchant and passenger ships as long as they were American built and crewed. This decision grew in part out of the vital role merchant ships had played as troop transports during World War I. By the end of the decade the United States Line, Presidents Line, Matson Line, Dollar Line, and other U.S. shipping and passenger fleets had become globally competitive, despite a decline in steerage business due to more restrictive U.S. immigration laws. American-flagged passenger ships were able to provide quality service across the North Atlantic, Pacific, and south to the Caribbean and Latin America, including small-ship service between Key West, Florida, and Havana.

At the same time, south Florida was becoming the target of a Roaring Twenties development boom that marked the first major landfill of coastal wetlands for nonagricultural purposes (it also marked the age of the land scam, as far more swamp acreage was sold than was actually filled in).

Henry Flagler opened up Florida's east coast by building a network of railroads and hotels including his famous "Flagler's folly" rail line linking the mainland and Key West, completed in 1912. Hundreds of workers died during its construction, including 300 men in a single hurricane in 1906. In 1935 another hurricane killed more than 400 retired World War I veterans and tourists being evacuated from the Keys as 50-foot seas swept their train off its elevated rail bridge. North of the Keys a dredged, terraced, and paved island known as Miami Beach was becoming renowned for its "millionaire's mile" of fabulous mansions.

It was during the flapper era that extended beach vacations became possible for significantly broader numbers of Americans. The expansion of middle-class culture saw the introduction of widespread travel advertising, more revealing ladies swimwear, the transformation of the suntan from a sign of outdoor labor to a symbol of beachfront leisure (a trend encouraged by Coco Chanel, among others), and, most important, the expansion of the U.S. highway system to coastal areas previously inaccessible by rail. One of the most popular roads built during this era was the Dixie Highway, running from Chicago to Miami, where automobile trailer parks began to make their appearance.

It seemed that nothing could slow this reckless new beachfront expansion, driven as it was by the burgeoning real estate, construction, and tourism industries. In 1928, the year state engineers completed the Tamiami Trail through the Everglades, connecting the east and west coasts of southern Florida, a hurricane killed 2400 Floridians. After the cleanup, they made a commitment to standing their ground and kept on building in harm's way.

Some scientists were beginning to warn of the vagaries of exposed coastal zones, tidal currents, maritime weather, even the living creatures of the sea. In the 1920s oceanography was still, from a practical point of view, a shore-based science, with few deep-water ships available for researchers beyond those temporarily on loan from the navy for hydrographic sea mapping and oil exploration (the navy was converting from coal to diesel fuel). Most oceanographers were out of the biological disciplines—zoology, botany, and ichthyology—and engaged in limited studies of tides and marine plants and animals. In 1903 the Berkeley zoologist William Ritter set up a marine biological station at a boathouse in San Diego with financial help from newspaper publisher E. W. Scripps and his sister, Ellen.

In 1925 Bill Ritter renamed his marine station the Scripps Institution of Oceanography. He believed the job of marine scientists was to look at whole organisms and their relationships to their environment, a field of study later defined as ecology. Up the coast in Monterey a wild and wildly innovative biologist named Edward F. Ricketts was also helping to introduce the idea of ecology to the study of the Blue Frontier. His book *Between Pacific Tides* (1939) emphasized the need to comprehend not only individual species but their interactions in the near-shore environment, where much of the ocean's productivity occurs. His collection of plants and animals from the kelp forests and submarine canyon, along with the work of Stanford University's Hopkins Marine Station, marked Monterey as a center for marine study. Ricketts's casual, party-friendly lifestyle also inspired the character Doc in the 1945 novel *Cannery Row,* written by his close friend, Nobel Prize–winning author John Steinbeck.

In 1930 the Rockefeller Foundation, under the guidance of long-time Rockefeller adviser Dr. Wickliffe Rose, decided to establish an East Coast oceanographic science center to match the West Coast work being carried out at Scripps and Hopkins. Rose was convinced that marine fisheries could play an important part in global agricultural and food production, provided enough scientific research was brought to bear. Along with new funding for Scripps and for ocean research at the University of Washington, $3 million was earmarked for the establishment of the Woods Hole Oceanographic Institution, to be located in the village of Woods Hole on Cape Cod, where a government fisheries lab and the Marine Biological Laboratory already existed.

Despite a few bright spots like the Rockefeller funding and the 1934 descent of biologist William Beebe and his friend Otis Barton half a mile below the sea in a 2.5-ton bathysphere that briefly inspired America, the Great Depression hit the waterfront hard. Fish landings and international shipping trade declined steeply as the national gross domestic product dropped by 25 percent. Among the hardest hit were merchant seamen and longshoremen. By 1933 an able-bodied seaman was earning $53 a month, down from $85 in 1920. But at least a seaman had steady if dangerous work. Longshoremen were earning around $40 a month, but only if they were able to survive the "shape-up," where crowds of desperate men gathered around a foreman who would pick a work crew for each ship to be loaded or offloaded. Often the only way to ensure a shot at a job was through bribery or strong-arm tactics.

As worker dissatisfaction grew, the International Longshoremen's Association of the American Federation of Labor began a union drive on the docks. Among its more effective organizers was a short, hawk-faced Australian named Harry Bridges, who worked the San Francisco waterfront. Years later one of his lawyers would remark that San Francisco had eight bridges and the one you didn't want to cross was Harry. On May 9, 1934, Bridges led the San Francisco longshoremen out on strike, demanding higher wages, shorter working hours, and a union hiring hall in place of the shape-up. Twelve thousand other longshoremen threw up strike pickets in Seattle, Tacoma, Portland, San Pedro, San Diego, and a dozen other West Coast ports.

By July 3, with the strike still going strong and frustrated shipping magnates calling for suppression of the "Red revolt," San Francisco's mayor decided to send the police down to the Embarcadero waterfront to "open up the port." This led to widespread rioting, with tear gas and bricks flying throughout the day. After a Fourth of July truce, the battle resumed on "bloody Thursday," when police opened fire with gas, pistols, and riot shotguns, killing two workers and wounding hundreds more. The governor then ordered in thousands of National Guard troops to secure the port, and for a time Bridges thought the fight was lost. But following a funeral for the dead workers, attended by 35,000 mourners, the Bay Area's unions organized a general strike in solidarity with the longshoremen. Close to 130,000 strikers shut down San Francisco and Oakland's stores, factories, and transportation systems. Sympathy strikes shut down ports around the country. Under pressure from the Franklin D. Roosevelt administration to settle, the shipowners agreed to recognize the longshoreman's union and meet its demands. Later that year, facing the threat of another strike, they recognized the Seaman's Union and its demands for eight-hour at-sea watches, an eight-hour day in port, and improved pay.

As U.S. maritime labor became increasingly organized and effective, a new wave of ship construction got underway, only it was not taking place in the United States but in Germany, Italy, and Japan. In 1936 President Roosevelt responded to this new maritime threat by expanding U.S. naval and merchant ship construction, calling it a public works project. The following year he appointed Boston political fixer and post–Prohibition financier Joseph Kennedy as chairman of the U.S. Maritime Commission with orders to speed up the shipbuilding effort. He tapped the right man for the job. "We are going to lay the keels for new fast ships. And we are going to do it now!" Kennedy pledged.

While Roosevelt saw the storm clouds of war and fascism rising in Europe and Asia and was convinced the United States would soon have to get involved, most Americans, including most military leaders, believed the United States could remain neutral and isolated from the conflict, protected as it was by two vast oceans.

One newly arrived immigrant knew better. Even though she was only 16 years old at the time, my mother, Eva Lieberg, had seen the face of fascism. At age nine she had been interrogated by the Gestapo in her home town of Ellrich Am Harz. Her mother, Emi, was not allowed to be with her at the time. Eva's father, Fritz, was taken off to Buchenwald concentration camp on November 9, 1938—Kristallnacht, the night of broken glass. There he was tortured for two months and put through a mock execution. About 250 of the men with him were murdered or committed suicide. In January 1939 he was let go. Many of the 30,000 Jews arrested on Kristallnacht were released at this time, only to be picked up with their families a few months later and murdered. In the spring of 1939, after increasingly frantic efforts and surrendering all their remaining cash to the Nazis, the Liebergs were able to make their escape to Holland. Fritz, Emi, and their three daughters were among the last German Jews to get out alive.

From Holland they were able to procure U.S. visas and book passage on the SS *Rotterdam* on its final trip before the Nazis invaded. The night before they were to leave, another liner, the *Simon Bolivar*, hit a mine and sank. Dozens of the girls' young schoolmates drowned. It took two weeks for the overcrowded *Rotterdam* and its seasick passengers to make it through the minefields and across the Atlantic to Hoboken, New Jersey. Once stateside, the Liebergs tried to arrange passage from Germany to Shanghai for Emi's older sister, Helene, but it was too late. She died in the gas chambers at Auschwitz. Her husband starved to death at the camp in Theresienstadt. Other relatives who had made it to England faced nightly bombing raids by the Luftwaffe. The Liebergs had no illusions about their new American home being safe, but they did believe it to be strong. That strength would first be challenged, not by Nazi Germany, but by its Axis ally, Japan, on December 7, 1941. That day of infamy would not only bring America into World War II but also fundamentally change how the United States related to its ocean frontier for the remainder of the twentieth century and into the twenty-first.

[Chapter Three] Oceanographers and Admirals

Only God Almighty and naval research can save us from the perils of the sea.

—SENATOR JOHN WARNER AT A NAVY DINNER

If I'd seen a Russian footprint down there instead of a fish, we'd probably still be down there.

—DON WALSH, RETIRED NAVY OFFICER AND ONE OF ONLY TWO HUMANS EVER TO GO TO THE DEEPEST POINT ON EARTH

U ntil quite recently few U.S. marine scientists and no major oceanographic centers took a leadership role in alerting the public to the environmental dangers facing America's Blue Frontier. This failure has little to do with any lack of insight on the part of scientists and their institutions and much to do with the hidden history of U.S. oceanography. It is a story of how science, which works best as a system of broad, open, and shared inquiry, can be distorted by single-source funding and state secrecy. It is a story that ironically has its origins in one of the most recorded moments in American history.

Several years ago I had the privilege of interviewing a number of veterans of the December 7, 1941, attack on Pearl Harbor. Dick Fisk was a 19-year-old marine bugler on board the battleship *West Virginia* that morning, watching what he and a friend thought was a U.S. Army plane making a practice run on their ship, launching its torpedo.

"We moved to the port side. We thought it was going to be a dummy torpedo until the explosion hit and then it seemed like this great big wall of water just washed us across the deck to the other side, which was 118 feet wide."

As the first wave of Japanese planes arrived over their targets, Commander Mitsuo Fuchida radioed a coded message, "Tora, Tora, Tora," to let the Japanese fleet know complete surprise had been achieved.

Joe Morgan was a third-class petty officer stationed on Ford Island that day. "When we ran outside to see what we thought was a plane crash, we saw a plane dropping bombs across the runway from our hangar in the area of the patrol wing squadron.

"As the plane pulled out we could see the meatballs [the Japanese rising sun symbols] under his wings. I knew then that we were really in war and I was afraid of being killed. About that time this plane came by flying slow, and I could see the goggles of the rear gunner. As he swung his gun around in our direction and started peppering our vicinity with machine gun bullets, I jumped behind a rear-wheel tractor for protection until after he passed.

"By that time it just seemed like bullets were coming from everywhere because planes were strafing from different directions and so I went into the end of the hangar there for protection and it sounded like hail on the roof.

"Here I was an aviation ordnance man that handled machine guns and knew how to operate them and I was hiding behind this I beam. So, my guilt overcame my fear, and I went ahead and started putting some machine guns in the planes that were sitting out on the warm-up mat. The PBYs had three gun mounts, so we put all of our guns that were available at that time in these planes and started shooting back at the Japanese.

"There was a plane that dropped his bombs over and around battleship row and was flying lowly across the landing field. As it started coming in our direction, we all, nearly every gun in our squadron, was shooting at him. As he passed over our hangar with everyone shooting at him, he burst into flame. And instead of crashing in the water off the shoreline there, he landed right on top of the USS *Curtis* on the crane deck, and the whole top of the ship then burst into flames and that was one of the most memorable sights I still remember.

"So I'm seeing things going on all around, mainly looking for planes. As I happen to be looking out toward the *Curtis* again, I saw this little submarine come to the surface, this midget submarine that came to the surface and was aiming at the *Curtis,* and so the *Curtis* gunner started shooting at him and put two 5-inch shells right through his conning tower, and I actually saw the holes appear in the conning tower. And then about that time the USS *Monihan,* the destroyer, came dashing down the channel, heading to get out of the harbor, and they rammed it head-on and then as they passed over dropped two depth charges on it. They had to be going so fast to keep from blowing off their own fantail in that shallow water, but they couldn't make the curve in the channel and they ran aground."

Within minutes of the attack local medical facilities began receiving heavy casualties. Dr. William Cooper, a civilian volunteer, arrived at Shipler Hospital, where he was introduced to the wartime practice of triage, choosing who should live and who should die.

"It was a shock to see how much bullets could tear a guy apart. The first case that was put on the table for me to take care of didn't have any legs and had one arm gone and the second arm was just hanging. . . . and I was going to start and give him some kind of assistance as he was still alive, but the triage gentleman said, 'No, take him off the table. Take him someplace else. We cannot take care of people like this. We only want you to help the ones who can fight tomorrow.'

"After working all night and most of the day I thought I'd go out and get something to eat. I was tired and as I came across one of the corridors of the hospital on about the second floor, there were glutted garbage cans. What in the world? Garbage cans up here? So I looked inside, and, my God, they were filled with arms and legs of people who had been hurt, damaged, and cut up. I said, 'My God, this is awful.'"

Dick Fisk, the marine bugler on board the *West Virginia,* had a brother working as a medic at Scofield Barracks.

"Frank worked for over 72 hours straight. He said he never saw so many wounded people in his life and dead bodies up there. He finally collapsed going into one of the operating rooms and his white uniform was completely covered with blood, and so they thought he was dead, so they put him on a gurney, and they moved him into the morgue. And my brother said he slept for over 24 hours. He said it was the best rest he ever got. He said nobody bothered him, except when he woke up he was awfully cold."

Battleship row ran along Ford Island for more than a mile. Of the nine battleships in Pearl Harbor that morning, seven were moored there, tied up to large white concrete piers known as dolphins. The *Nevada* was the only one to get under way during the attack, although it failed to reach the open sea.

Next came the *Arizona* moored next to the repair ship *Vestal.* The battleship *West Virginia* was behind it, moored outboard from the *Tennessee.* Behind them were the *Maryland,* the *Oklahoma,* which later capsized, and the *California.* After recovering from the shock of the first torpedo to hit his ship, Fisk resumed his battle station on the bridge of the *West Virginia.*

"We took a total of eight torpedoes actually and four bombs and then it was about twenty minutes after eight and one of the bombs came down. . . . We didn't see the shrapnel come over, but the next thing we knew, we saw Captain Bingham hit the deck. He let out a big yell and we looked over to him and most of his stomach was missing. He had a terrific hole in there. By the time they got him down to the boat deck, which was about 11 or 12 minutes later, he finally died.

"We fought fires and people were coming up on the decks. Most of them had so much oil stuck to them that they were on fire, and we were trying to roll on top of them to put the fire out. We stayed aboard ship fighting fires. And even then the bombing, the strafing was still going on, and you could hear the machine-gun bullets whiz past you, but nobody seemed to pay any attention to it. I guess maybe we were in shock but we didn't pay much attention to that. We were just interested in getting our friends and buddies squared away.

"We saw this big bomb come down, and we all thought it was going to hit us. We were hurled up against the forepart of the navigation bridge, from the concussion of it, and then we got back up on our feet and we looked back, and we just saw the *Arizona* and it was just one tremendous ball of fire. It was the most devastating thing I've ever seen in my life. I never realized that a ship could blow up like that so fast. The bow just came completely out of the water, and when she settled down she was just one great big fireball, and we didn't realize it then but she had lost over 1177. Right after the fireball I looked, and I remembered two of my friends that I went through high school with, James and Finley, and I was thinking to myself, 'God, I hope they're all right,' and come to find out later that they were killed in that explosion."

The explosion on board the *Arizona* that day, caused by a bomb striking its forward magazine, was the largest single detonation of World War II until atomic bombs were dropped on Hiroshima and Nagasaki.

"There's guys swimming by the *Arizona* and we could hear them crying, 'Help, help, help. I can't see.' There was quite a few of the guys that had so much oil stuck in their eyes that they were kind of swimming in circles. So we dove in the water and we brought all of those guys out, the wounded, and laid them along the shoreline. And then we went back in the water to get all the bodies."

Twenty-four hundred Americans died at Pearl Harbor on December 7, along with some 130 Japanese submariners and airmen. Of the 1543 men aboard the *Arizona* that morning, 1177 remain entombed in its wreckage, now a memorial. They represent every rank in the United States Navy, from seaman apprentice to rear admiral.

The Japanese attack on Pearl Harbor was coordinated with military offensives against Malaysia, Hong Kong, Guam, Wake, Midway, and the Philippines. Militarily, the Pearl Harbor attack was a great but not unqualified success. Although the fleet was shattered, America's aircraft carriers were at sea and escaped destruction and would soon hammer the Japanese fleet. The failure to hit Pearl's power plant and fuel supplies allowed the U.S. Navy to rebuild quickly. But more important than military considerations, the Japanese ruling circles completely misjudged the American public's response to the attack. Rather than undermine the U.S. will to fight, December 7 unified a divided nation in its commitment to total war against Japan and its fascist partners.

The day after the attack, President Franklin Delano Roosevelt went on national radio and declared: "Yesterday, December 7, 1941—a date which will live in infamy—the United States of America was suddenly and deliberately attacked by naval and air forces of the empire of Japan. . . . No matter how long it may take us to overcome this premeditated invasion, the American people in their righteous might will win through to absolute victory."

In the wake of Pearl Harbor, the nation rapidly mobilized its military and civilian resources. At Woods Hole, Scripps, and other marine stations, scientists joined the war effort. Scripps helped organize a University Division of War Research at the navy submarine base on Point Loma in San Diego, which was overseen by Scripps geologist and navy lieutenant Roger Revelle, a tall, personable man who would soon become the first oceanographer of the navy. Among his colleagues were the physicist Carl Eckart and a young Austrian-born graduate student named Walter Munk.

Until then marine research focused primarily on the sea's living resources—fish, invertebrates, marine mammals, and their interactions in the water and along the shore. "Now, physical oceanography, the study of acoustics, water temperature, currents, the bottom structure, visibility, all things that might affect submarine or ship operations, came into their own," recalls Walter Munk.

While Revelle's people were working on acoustics for sonar, Munk and Scripps director Harald Sverdrup taught the navy how to predict swell and surf conditions for amphibious landings in North Africa, Italy, and Normandy.

On the East Coast, Woods Hole director Columbus Iselin and investigator Maurice Ewing discovered how underwater sound location was influenced by temperature and salinity, identified the sonar interference caused by snapping shrimp (called the deep scattering layer), and found deep sound channels within the sea, all major breakthroughs for submarine war fare. One of Iselin's students, future Woods Hole director Paul Fye, worked on underwater explosives, while Allyn Vine and other scientists developed portable depth and temperature devices, underwater cameras, antifouling paints, and improved weather forecasting that helped the navy win the war at sea.

Late in the war Revelle, now oceanographer for the navy's Bureau of Ships, was traveling frequently between the Pacific theater, California, Woods Hole, and Washington, D.C. In 1944, seeing the promise of postwar work for his colleagues, he set in motion what became the Scripps Marine Physical Laboratory (MPL). The MPL was an acoustics lab that functioned as part of the university but also conducted top-secret antisubmarine warfare research for the navy. Other navy-funded university weapons labs included the applied physics labs of the University of Washington, Johns Hopkins University, Pennsylvania State University, and the University of Texas at Austin.

The end of World War II quickly segued into the Cold War with the Soviet Union and the testing of U.S. atomic weapons in the Pacific. In the winter of 1945, Revelle was put in charge of oceanographic studies for Operation Crossroads, a test firing of two nuclear weapons against navy ships at Bikini, an atoll in the Marshall Islands.

"When the task force arrived, a lot of people from Scripps and Woods Hole came out. We had most of the oceanographers in the country out there. There weren't so many in those days," Revelle half smiled, recalling the scene for me some 35 years later. "The most striking event, the most vivid event that nobody really expected, which was quite frightening, was this phenomenon called the bay surge from the underwater explosion—the second test. All the ships had been moved outside the lagoon except the target ships, and we were watching and saw what looked like a huge wave about 100 or 150 feet high moving rapidly out from the center of the explosion toward the beach. This thing only went part way and then stopped, and then it lifted off the water. You see, it really wasn't a wave at all. It was just a cloud of mist and spray about a mile across that lifted from the lagoon. Then it began to rain on us from this cloud. . . . All the ships got badly covered with radioactive water. For days thereafter we had the crews trying to clean up the ships. A lot of the radioactivity stuck to the paint. We never really did get any of the ships decontaminated."

In August 1946, President Harry Truman signed a bill establishing the Office of Naval Research (ONR). The original idea was that this would be the nuclear propulsion center for the navy. The ONR quickly began funding university-built nuclear accelerators, but the Bureau of Ships seized the initiative to develop nuclear propulsion for the navy, handing it off to a young lieutenant named Hyman Rickover. (Rickover went on to become a highly efficient but autocratic admiral, the J. Edgar Hoover of the nuclear navy.) With its original purpose undermined, the ONR shifted focus to wider support for basic science that might be of interest to the navy. Under Vice Admiral Harold Bowen, Revelle became director of the ONR's Geophysics Branch, which oversaw oceanography, and helped establish the contract language for its university research grants. In 1948 he left the ONR to help run Scripps and was replaced by the geophysicist Gordon Lill, who oversaw the transformation of marine science from a living-resources orientation to a physical-engineering perspective. By 1949 the ONR was spending more than 75 percent of its basic science budget in the physical sciences and only 15 percent in biological and medical sciences. "The value of research in nuclear physics is more obvious to the navy than is the value of research in the life sciences," noted *Scientific American* magazine in February 1949.

The ONR's $43 million in contracts in 1949 represented 40 percent of the nation's total science spending. It was providing millions of dollars and surplus ships to the navy's friends at Woods Hole (Iselin), Scripps (Revelle and Eckart), and Columbia University's Lamont Geological Observatory (established by Maurice Ewing). Lill also began seeding the coasts with new ocean centers, creating or expanding programs at the University of Washington, University of Miami, Oregon State, Texas A&M, and the University of Rhode Island, where he handpicked John Knauss to direct the program (Knauss, who studied under Walter Munk, would later head NOAA under President George Bush).

"We were the major player in determining what went on with oceanography. It was fun being there," recalls Art Maxwell, a former navy man and grad student under Revelle who joined ONR as Lill's assistant and later succeeded him.

"By the 1950s Gordon Lill is in effect the navy czar of civilian oceanography," says Gary Weir, chief of the Contemporary History Branch at the U.S. Naval Historical Center. "A generation of scientists now depend on their friends in the navy to keep them afloat, Woods Hole more than the others [since it has no university funding].

"Because the research is navy driven, biological oceanography and biology are neglected," Weir continues. "Physical oceanography and geology and chemistry are brought to the fore. The heads of the oceanographic institutions are aware this is a problem, but they also know they have to pay a price for what they're getting and that price is to place the emphasis where the navy wants it to be. . . . If you're Columbus Iselin at Woods Hole you know 70 percent of your budget comes from ONR. You're not going to lie to Gordon Lill, and if you do he'll know it because you've been friends for years and drink together and go out to sea together."

Although Congress established the National Science Foundation in 1950 as an alternative source of research funding (under the leadership of ONR Chief Scientist Alan Waterman), the navy continued to provide two-thirds of all federal spending on marine science into the 1970s.

"The job of people like Lill and myself was to convince the rest of the navy that anything we supported in oceanography was also good for the navy," recalls Art Maxwell. That job was made easier after civilian oceanographers helped the navy get into the nuclear weapons game.

In the early days of the Cold War, the army and navy saw themselves confronted by an enemy that terrified them, making them fear for their very survival. This was, of course, the Air Force with its monopoly on nuclear bombers and ongoing work developing long-range ballistic missiles (using former Nazi rocket scientists like Wernher von Braun).

In 1949 Secretary of Defense Louis Johnson came out in support of the procurement of B-36 long-range bombers capable of dropping nuclear weapons on as many as 70 Russian cities. "B-36 Can Blast 70 Red Bases, AF Says," was the headline the *Washington Daily News* ran after classified documents were leaked to UPI. At the same time, Johnson rejected navy plans for a "supercarrier" capable of launching long-range atomic bomb toting aircraft. Shortly thereafter, Congress received an anonymous document claiming funding irregularities in the B-36 program. This led to a series of hearings at which high-ranking navy officials complained to Congress about having their feelings hurt. The press dubbed this the "revolt of the admirals."

The Air Force's bombers and growing missile program continued to represent America's primary nuclear strike force throughout the 1950s. After the Soviet Union developed its own bombs, America's first strike plans evolved into a strategy called MAD, for mutually assured destruction. During this period the army and navy were reduced to working on tactical nukes—atomic land mines, artillery shells, torpedoes, and depth charges.

In 1955 Roger Revelle and the MPL's chief scientist, Al Focke, chose a deep-water site 450 miles southwest of San Diego for Operation Wig-Wam, the navy's test-firing of a nuclear depth charge. The blast was not supposed to break the surface, but it did. Two days later, on May 16, 1955, a spike of radiation passed over San Diego.

"Those bomb tests made an antiweapons man out of me," Revelle later admitted. "I feel very strongly about it. I don't think anybody should be developing nuclear weapons. Believe me, no one who's seen one of those things go off ever wants to see another one."

Still, in 1956 Woods Hole hosted the Undersea Warfare Committee of the National Academy of Sciences. Dubbed Project Nobska (after a local lighthouse), the summer-long retreat was held at the Whitney estate and included the nation's leading oceanographers and some outside scientists like Edward Teller, "Father of the H-bomb." Out of this woodsy idyll came three volumes of proposals for new approaches to naval warfare, including the idea of firing nuclear-tipped missiles from submerged submarines, using small warheads, solid-rocket fuels, and advanced guidance systems. By 1960 the submarine *George Washington* had test-fired a Polaris missile at sea, and the navy was able to claim the third leg of what became known as America's strategic nuclear triad of Air Force bombers and missiles and navy "boomers," or missile submarines. The next 30 years would be dedicated to protecting the boomers, while going after Soviet subs, using satellites, helicopters, surface ships, and, of course, other submarines.

The civilian oceanographic centers were by 1960 top-heavy with physical oceanographers and physicists with high-level security clearances and big-money contracts not only from the navy but also from the Air Force, Central Intelligence Agency, and DARPA (the Pentagon's Defense Advanced Research Projects Agency). "The physical oceanographers became the directors of all the stations and the marine biologists felt like they were now second-class citizens," explains Deborah Day, senior archivist at Scripps.

The growth of physical oceanography made some significant contributions to the earth sciences by establishing the theory of plate tectonics (continental drift), based on the work of Walter Pittman, Bruce Heezen, and Marie Tharp at the Lamont Geological Observatory and Harry Hess at Princeton. Roger Revelle also made a major contribution by linking rising levels of carbon dioxide in the atmosphere to industrial greenhouse-gas emissions from the burning of fossil fuels. Among those who would take note of his work was a student at one of his Harvard seminars named Albert Gore, Jr.

The institutes were also blessed with a growing fleet of navy-funded blue-water research ships and vessels including Woods Hole's *Alvin* (named after Al Vine). Delivered in 1964, *Alvin* was a three-person submersible capable of diving more than a mile down. In 1966 it was used to locate an H-bomb lost off the coast of Spain (after a B-52 collided with its air tanker during refueling). Among Woods Hole's navy liaisons was a young ONR ensign Bob Ballard, who would later use *Alvin* to locate the wreck of the *Titanic*.

Unfortunately, oceanographic biology, marine ecology, and work in the nearshore environment were badly neglected during the Cold War. One exception was the California Cooperative Oceanic Fisheries Investigations (CalCOFI), funded by the state of California after the 1947 collapse of the

sardine fishery. For more than half a century CalCOFI continued to provide ongoing data on the California current and its living resources.

In 1951 Rachel Carson, then an employee of the U.S. Fish and Wildlife
Service, also managed to break through the martial drumbeat of the times
with publication of *The Sea Around Us.* This natural history of the oceans
sold more than a million copies, introducing the public to what were then
obscure concepts like the marine food chain, ecosystems, and the biosphere.
It inspired an Academy Award–winning documentary and a follow-up
book, *At the Edge of the Sea.* In 1962 Carson shifted focus to the dangers
posed by DDT and other pesticides with her most famous book, *Silent
Spring,* which in turn provided a major boost and inspiration to the growth
of environmental protest in the United States.

In 1970, during the height of the Vietnam War and protests against the
war, Mike Mansfield, the Democratic House majority leader, pushed
through an amendment to a military spending bill that limited ONR funding to work that had direct application to the navy's mission (to help win an
unwinnable war). This marked a transition point, after which the National
Science Foundation became the major funder of oceanographic research.
Still, the priorities of the field would remain in the hands of physical
oceanographers and military physicists, people like longtime Scripps director Bill Nierenberg, a former assistant secretary general of NATO, and
Woods Hole director Craig Dorman, a retired navy admiral in charge of
antisubmarine warfare. The navy also retained a monopoly of funding in
two areas it saw as critical—acoustics and, oddly, marine mammals.

"After Maurice Ewing discovered the deep sound channel, the navy
decided this was so important that acoustics kind of went behind a curtain
for twenty years and the ocean community had little to do with the acoustic
community," says Walter Munk. "In retrospect, we would have been better
off without such heavy security."

While the navy was using its SOSUS underwater hydrophones to listen in
on Russian submarines, the audio "garbage" it was trying to filter out proved
to be a treasure trove for discerning scientists with top-secret clearances.

"You look at when SOSUS was opened up [for civilian use] in 1993. All
these scientists were surprised that the number of marine mammal sounds
and underwater volcanoes were a magnitude of order greater than they'd
been recording," says Navy historian Weir. "All except people like Munk,
who had been using it for years."

"We had a kind of treaty with ONR, where if we didn't identify the
location or aperture [of the SOSUS arrays], we could use the information
we collected," Munk recalls. "Still, some science journals turned our work

down, saying if you can't identify your sources of information, we won't publish your work."

This had to be frustrating for scientists with access to the navy's secret sound systems, satellites, submarines, and sea maps, like being allowed to take an open-book exam on the oceans while everyone else has to guess the answers, but then not getting to see your grades posted on the bulletin board.

Former Woods Hole geologist (and navy commander) Ballard used his access to the navy's top-secret tools to realize a lifelong ambition. According to an article by Laurence Gonzales in *National Geographic Adventure* magazine, Ballard used the NR-1, the navy's miniature 10-man nuclear spy submarine, to survey the wreckage of the USS *Thresher* back in 1984. This gave him the key clue he needed to locate the *Titanic*.

"Suddenly it all became clear," Gonzales wrote of Ballard. "To find a deepwater wreck, you don't look for a hull a few hundred feet long. You look for a trail of debris a mile or more long. And when you find it, it will take you directly to the heaviest debris, the ship itself."

Since Ballard discovered this debris scattering effect (reflecting the difference between crush depth and bottom depth) on a top-secret navy mission, he was able to keep the technique to himself. The knowledge came in handy a year later, in 1985, when he used the same approach to locate the wreck of the *Titanic* in an area of the North Atlantic where numerous other search efforts had failed.

"No Dogs Allowed Including Military Working Dogs. Code 35 Authorized Persons Only. Code 35 Marine Mammal Program," reads the sign on the gate inside the San Diego sub base on Point Loma. It's a warm fall day with the temperature in the low eighties as longtime navy public affairs man Tom LaPuzza guides me onto the wooden pier that smells of creosote and fish. Open to San Diego Bay are a series of 30-foot-by-30-foot enclosures with dozens of dolphins and a pair of white Beluga whales from the Arctic swimming about in their own double-wide enclosure. A few pelicans and a snowy egret are hanging around as trainers feed fish from a bucket to a couple of their charges. A monarch butterfly flits by. The dolphin closest to me is lolling on the surface, looking, if I may anthropomorphize, bored silly. Just beyond the dock a Boston whaler is pulling up next to one of the corrals.

"We're going to tone test some of the dolphins before exposing them to noise generators," LaPuzza explains. "Then we're going to retest them for tone response to see if they've experienced any temporary hearing loss."

"What?" He looks to see if I'm making a joke. I realize this is about active sonar.

The navy plans to deploy a $350 million sonar system that would project 215 decibels of sound into the water and bounce it off enemy submarines (Iraqi, North Korean, whatever). Environmental groups believe that the underwater noise could prove harmful to marine mammals. The navy's dolphins have been volunteered to test this theory. Initial testing suggests that it is a problem in shallow waters. In 1997 and again in 2000 beaked whales stranded themselves following navy and NATO testing of active sonar systems of more than 200 decibels in the Bahamas and the Mediterranean. The navy denied any connection. Still, the navy isn't spending $20 million a year on the care, feeding, and transport of 75 dolphins, 12 sea lions, and two beluga whales in order to test their hearing.

Many critics, from animal rights activists to congressional budget hawks, wonder why they are spending the money. It is hard to grasp the extent to which the navy has been involved with marine mammals since the 1960s. Through ONR it has employed or provided grants to almost every major scientist involved in work on dolphins and sea lions including the late *Man and Dolphin* author and popularizer of dolphin intelligence John C. Lilly. It has worked closely and traded animals with a number of schools and ocean theme parks, including the University of California at Santa Cruz, the University of Hawaii, the New England Aquarium, Sea World, and various dolphin petting zoos in the Florida Keys. Even the *Sierra Club Handbook of Whales and Dolphins* had its origin as a navy technical report. The Russian navy also used to own dolphins but shut down its program in 1998 when it realized it barely had enough mullet to feed its sailors.

The U.S. Navy's original interest in dolphins was based on the short-range high-resolution sonar and hydrodynamics of the animals, but researchers soon began viewing these highly sociable animals as potential seagoing K-9s.

"The fleet animals are the operational animals," LaPuzza tells me. "We've used them in mine hunting exercises around the world. They're also very effective at swimmer defense. When you send a dolphin out, the dolphin tells our guys there's a swimmer in the area, then the dolphin is removed from the scene and the guys deal with it."

"The Gulf was their last combat deployment?" I ask.

"Nineteen eighty-seven to 1988 off Bahrain. A swimmer defense system was used around the command ship *Lasalle*."

This was during the Iran-Iraq War, when the U.S. Navy was escorting oil tankers through the Persian Gulf, and the Reagan-Bush administration was providing satellite intelligence to Saddam Hussein.

"At Bahrain a commander drinking at a party said the killer dolphins are coming, and no one swam in that harbor again, so they certainly have a deterrent value," LaPuzza grins. "There never really were killer dolphins in Vietnam or elsewhere. That's a lie."

Which is not the story I've been told.

In 1971 six navy dolphins were flown from Kaneohe Bay, Hawaii, to Cam Ranh Bay, Vietnam, where enemy divers were sabotaging U.S. ships. Once in-country, the dolphins were housed in an elaborate flotation craft built in San Diego. A dolphin would scan the water in the bay with its sonar. If the water was clear, the animal would push a normal response paddle. If it sensed an intruder, it would hit a second paddle that opened its pen gate. Here's where the stories diverge. According to the navy, the dolphin would then clamp a marker onto the swimmer and alert its trainers.

According to former navy and CIA dolphin specialist Michael Greenwood, who gave 150 pages of closed-door testimony to Frank Church's Senate Intelligence Committee in 1975, the dolphin would run its beak into a padded cone placed in a water-level weapons rack. The cone's tip contained a heavy-gauge hollow-point needle attached to a carbon dioxide cartridge, a weapon developed by the navy as an antishark device. The dolphin would then swim to the diver and plunge the needle into him. The carbon dioxide gas would explode the diver's organs from within, floating the corpse to the surface for recovery by the dolphin's trainers. This was known as "swimmer nullification."

Nineteen seventy-seven articles in *Penthouse* and the *Christian Science Monitor* claimed that the navy's dolphins killed several dozen Vietnamese divers and two Americans who strayed into restricted water. James Fitzgerald, who helped develop the swimmer nullification program for the CIA, told *Parade* magazine that the navy's dolphins "blew up" a number of frogmen. In 1984 I interviewed a retired navy scientist who confirmed that killings took place during the early part of the dolphins' deployment but said that later on the dolphins made "live captures." He requested anonymity saying he didn't want to "give the navy a black eye."

Ken Woodal, a former navy SEAL I interviewed in Fort Smith, Arkansas, said he worked six months as an underwater demolitions man with dolphins in Vietnam, carrying out top-secret raids against North Vietnamese harbor facilities. The animals he worked with were not used for swimmer nullification but to place mines.

"The three dolphins I worked with were quite effective in attaching light mines to enemy wharves and piers," he told me. "We'd transport them from Cam Ranh Bay by C-130 aircraft and river patrol boats. They were never sacrificed. They were trained to detach from their mines and return to their pickup boats. The thing that still sticks with me after all these years is the intelligence of those water mammals. You'd listen to them in their pens or working together in the water and swear they were talking to each other."

Later, during the Reagan administration's defense buildup, the navy program expanded to some 140 animals. "We got a national defense exception to the Marine Mammal Protection Act that allowed us to capture wild animals [in the Mississippi Gulf]," LaPuzza says.

Even as plans were being made to deploy animals to guard Trident nuclear missile submarines in Bangor, Washington; Savannah, Georgia; Groton, Connecticut, and elsewhere, a few trainers began questioning the safety and reliability of the program.

"When I came into the program in 1985 they'd updated the dolphins' nose cones with a .45-caliber bullet ejector, like an [antishark] bang stick," says former trainer Rick Trout. "They'd practice ramming divers and dummies in the water in order to 'nullify' them."

In 1987 Trout participated in a weeklong nighttime dress rehearsal for the dolphins being deployed to the Persian Gulf. "I volunteered to play an enemy diver that they were supposed to detect, find, and nullify, only nothing went right. Mostly they'd run off to play with the wild dolphins in San Diego harbor or else spook and return to their pens and refuse to come back out. Finally, Toad, a female, used her sonar to locate me in about 20 feet of black water. Only as she approached she tossed her weapon, swam up to me, and lay her chin on my shoulder. The next day I went up to the program director and said, 'This doesn't work anymore, does it?' He said, 'We know that.' 'Then why are we here?' I asked. 'It's just a deterrent,' he said." After talking to reporters in 1988, Trout was fired and threatened with federal prosecution for revealing classified information.

In 1990 new charges were made by another trainer, David Reames, who had worked with the navy's sea lions. Reames claimed that the animals often failed to respond to trainer commands. While trying to maintain his security oath, he gave several examples of animals bumping "potentially dangerous classified objects" (the .45-caliber weapons) against scuba tanks, trainers, and boats. On at least one occasion when an armed seal jumped into a training boat, its human handlers jumped into the water to get away from it.

Today the navy insists its marine mammals are highly effective operatives for mine detection, surveillance, and security. But General Norman Schwarzkopf declined to have dolphins begging fish in his combat zone during the 1991 Gulf War, and no dolphins were deployed to the waters off Panama in 1989, Somalia in 1992, Haiti in 1994, or the Adriatic during the 1999 war in Yugoslavia.

Still, these last military draftees remain "operational," in large measure because of the navy's expanded interest in mine warfare. This includes what the navy calls its "Very Shallow Water/Mine Countermeasures" detachment of navy SEALs, marines, and dolphins working together to scout out and clear amphibious landing zones.

It is one of an array of new tactics and strategies being developed for work in the littoral regions of the world. The *American Heritage Dictionary* defines "littoral" as referring to a coastal region or shore. Throughout the Cold War, the navy (and U.S. ocean science) focused on the open seas or "blue water," where Soviet nuclear submarines were seen as the major threat to U.S. national security. With the loss of this global maritime threat and the beginning of base closures and defense budget cutbacks in the early 1990s, the navy and Marine Corps quickly found a new strategic justification for their existence and funding.

In September 1992 the navy and Marine Corps published *From the Sea*, followed two years later by *Forward . . . From the Sea*, arguing that they must be ready to project power from the continental shallows or "brown waters" of the world onto any shoreline where regional conflict might require or inspire intervention by the world's last superpower.

"We spent a lot of time looking at the deep ocean because that's where the major threat was," says Rear Admiral Paul Gaffney, head of ONR from 1996 to 2000. "Now with the increase in mine warfare and diesel subs, smaller quieter subs that can be obtained by Third World countries or any number of nations, you get into the shallows and it's a more complex environment. On the deep ocean bottom, on the abyssal plain, processes tend to be very gradual. The shallows by contrast change quickly. You can see differences taking place a meter apart and to try and predict them is very difficult."

Which is one reason the admiral remained a strong backer of continued funding in basic oceanographic research, even as Congress continued to hand him buckets of money for applied military research. ONR's 2000 budget was around $1.5 billion, of which Gaffney was able to sequester $400 million for broad-ranging basic research.

"Really I can't imagine an ocean topic not of use to the navy and Marine Corps," he tells me. "For example, we want to characterize maritime environments; reefs, seagrasses, barrier islands, mangroves. . . . We want to understand the processes taking place in these different [nearshore] environments so if we're going [to war] somewhere, and if there is not a comprehensive oceanographic survey that's been done [of that location], we'll still be able to see what's going on there, we'll have a reliable model of what takes place in that environment. That's our goal."

I can see why this charismatic officer has won the admiration of so many marine scientists, even those not applying to him for grants. His friendly brown eyes and beagle-like cheeks give him an engaging demeanor that counterbalances what might otherwise be an intimidating intellectual fleetness of mind as he pursues what he refers to as his "vocation," seeking to understand the Blue Frontier for America's maritime warriors. Since his days at the Naval Meteorology and Oceanography Command in Stennis, Mississippi, he has worked hard to declassify navy data that might be useful to civilian scientists, policymakers, and the public, while pursuing research partnerships with a number of other federal agencies. And, of course, the purse strings he's controlled look like massive towing cables to civilian researchers.

"Watch Gaffney at a meeting and see all the people form in his wake like remoras following a shark," suggested one observer of the phenomenon.

The admiral takes a sheet of paper, and folds it into three parts. He reopens it and writes "land" on the first third and "ocean" across the other two thirds. "This is pretty close to how the planet's divided," he explains. He then draws a thin double line along the land's edge, shading it in. "We're now six billion people and 50 percent of us live within 50 miles of an ocean. In our lifetime we've seen twice as many people living on the coast. And this creates environmental problems of pollution and this creates conflicts. And if the sea-level rise increases that will also affect people."

I mention that half the world's people are also living in cities for the first time in human history.

"Yes, urban warfare becomes more important," he agrees.

In 1999 the navy and marines staged urban warfare exercises in Monterey and Alameda, California. The first of the exercises, Operation Sea Dragon in Monterey, was scaled back after the California Coastal Commission voted 7–0 against an amphibious assault that would have landed helicopters and hovercraft on the beaches of the National Marine Sanctuary. The commission said the war game threatened endangered species including snowy plovers, sea otters, and gray whales.

"We need to understand what's going on here." Gaffney runs his finger along the thin line dividing paper land from paper sea. "This is where the navy and Marine Corps operate, in this area that's also under the most stress."

I think of America's long neglected frontier waters, and don't feel reassured. Our greatest future threats, I suspect, will come not from other guys with guns but oceanic horrors of our own creation.

[Chapter Four] Quarrel on the Littoral

Here he lies where he longed to be;
Home is the sailor, home from the sea,
And the hunter home from the hill.

—Robert Louis Stevenson, "Requiem"

When the Cold War ended, America did not demobilize, as it
had after previous great conflicts. . . . It remains configured
and equipped to confront a war of maximum scale, prompted
by some large and unknowable threat that no one can yet name.

—William Greider, Fortress America, 1998

Some 300 elegantly attired people have gathered under the star-spangled atrium of the cavernous Ronald Reagan Building in Washington, D.C., for the privately funded concluding gala for the "Oceanographic Greats" symposiums. The symposiums have been funded by the Office of Naval Research to the tune of $825,000. It has been a year-long millennial celebration of ONR's first half century and the people it helped to fund. The half-dozen seminars held at major oceanographic institutes around the country and the taping of some 40 oral histories were organized by the Heinz Center, named after the late Senator John Heinz of Pennsylvania, and presided over by his widow, Teresa Heinz, now married to Senator John Kerry of Massachusetts. Walter Munk, Bill Nierenberg, former chief of naval operations Admiral James Watkins, and many other old war horses are here at the gala along with a newer generation of naval officers and scientists. Congratulations have been sent from two former presidents, and a congressional resolution was passed applauding everyone's commitment and dedication.

After an hour of conversation, champagne, and canapés, several of the liveried waiters begin sounding the chimes for dinner. I find my name card at one of the more than 20 round tables at the bottom of a wide marble stairway, not too far from the podium. Senator John Warner of Virginia, a veteran of both the navy and Elizabeth Taylor, is chatting amiably with several people at the next table. I try to slip by. "And what do you do?" he asks, catching my eye. "I'm writing a book on the oceans called *Blue Frontier*." "It really is a frontier, isn't it?" he interrupts, patting me on the back. "And," he adds, "this night is very moving for me."

I take my seat. Soon we all rise for the national anthem and parading of the flag by a four-man navy color guard. I can't help but notice that the African American sailor in the color guard seems to be the only person of color among the hundreds of people here, not counting our Ecuadorian waiter, some other servers, and a few security guards lingering in the balconies.

There's a nondenominational prayer by a navy chaplain. The assistant secretary of the navy says a few words. We eat sautéed salmon and tenderloin of beef with morel sauce (not quite surf and turf). I talk with my tablemate John DeLaney, the scientist and explorer from the University of Washington who led the Juan de Fuca expedition. Teresa Heinz gets up and talks about her late husband's concern for protecting the coasts. Admiral Gaffney, in a military tux with gold chevrons on his sleeves, talks about the importance of the oceans to our way of life. He also mentions Heinz's generosity, presenting Teresa with a navy public service award. The assistant secretary pins it to her puffy-sleeved black gown. Senator Warner, chair of the Armed Services Committee, introduces himself as Petty Officer John Warner, Radioman 3d class, to appreciative laughter and applause. He jokes about the "War of Northern Aggression" and how the Yankee admiral Farragut said something about "Damn the torpedoes, I'm going to the head." Soon-to-be-defeated Senator Chuck Robb, also from Virginia, says a few uninspired words. He's followed by John Kerry, who jokes that as a navy man he was keeping a close eye on the assistant secretary as he pinned his wife. He also talks about nonpartisan support for the oceans, using as an example how he was really the one who wrote the 1996 Fisheries Reform Act, for which Republican Senator Ted Stevens received credit. Republican Representative Curt Weldon of Pennsylvania tells us that John Heinz was his mentor and taught him the importance of protecting marine ecosystems. Admiral Gaffney gets back up again, raising his glass. "I'd like to make a toast. The proper response is 'Here! Here!'" He pauses. "To Mrs. Heinz, scientists, students, and the U.S. Navy and the great maritime nation she serves."

"Here! Here!" the crowd responds. He bids us good night, good sailing, and fair winds.

I have nothing against people patting themselves on the back. If nothing else, it's a way of ensuring flexibility as one ages. Still, more than a decade after the end of the Cold War, not a lot has changed in the way the navy and the defense and intelligence establishments view the ocean and their role in it.

In 1991, following the collapse of the Soviet Union (which the CIA failed to predict), Republican CIA chief Robert Gates and Democratic Senator Al Gore of Tennessee consulted about a possible new role for the spy agency, studying global environmental change. One approach they agreed on was to start declassifying intelligence data useful for research in areas such as climate change and the collapse of marine ecosystems. A special task force was established to oversee MEDEA (Measurement of Earth Data for Environmental Analysis). Its 35 to 40 members would include people from the intelligence community, the navy, and civilian scientists with higher than top-secret security clearances.

Linda Zall has piercing blue eyes, straw-colored flyaway hair, a preference for bright red lipstick, and a kind of nervous animated personality (at least in the presence of a reporter). She is MEDEA's executive director and also runs the CIA's Environmental Center.

Established in 1997, the CIA's ecocenter is housed in the directorate of intelligence, and assesses the political and economic impacts that environmental issues have on U.S. security interests. Specific programs include assessing transboundary environmental crime (the smuggling of animal parts now ranks as the world's second largest black market, after drugs but ahead of gunrunning) and supporting treaty negotiations and assessing foreign environmental policies (European and Indian opposition to genetically modified foods from U.S. corporations, for example). The center also assesses the role played by the environment in regional instability and conflict, which can range from monitoring the environmental consequences of a NATO bombing campaign to looking at how water rights impact Israeli-Syrian negotiations over the Golan Heights.

Zall chose the MEDEA acronym to honor Medea, the mythic princess and sorceress who helped Jason and the Argonauts steal the Golden Fleece. JASON is also the acronym for a long-established science group that advises the Pentagon on its weapons systems. Of course the mythical Medea also bore Jason children and then killed them to avenge his infidelity. And that was before the CIA was even operational.

During the 1990s MEDEA began providing science institutes some limited information based on navy oceanographic data and overhead Department of Defense and National Reconnaissance Office (NRO) spy satellites. Still, only minimal resources were made available, according to Representative Curt Weldon, who sits on the House Armed Services Committee.

"I'm unhappy with the pace of it [declassification]. . . . Linda's having some financial setbacks now," the representative from Pennsylvania tells me during an interview in March 2000. "There's also the issue of trust and polarization from the past. In the past you were either pro-environment or you were pro-defense, but I see a natural linkage. We can do things for the ocean's benefit with spin-offs from defense." His aide interrupts him to say there's a vote in two minutes. Weldon excuses himself, leaving for a House floor vote on a $12.7 billion emergency appropriation, the centerpiece of which is more military aid to Colombia. I check out the models of different weapons systems on top of his office TV—an A-1 Abrams tank, a fighter jet, an Osprey tilt-rotor aircraft, a pilotless spy plane.

"I can take ocean research into the defense agenda," he tells me on his return. "They don't see me as a threat so I can convince them to do more. But I also blame the ocean research community for not doing enough. They think this [information and technology transfer] should take place just because it makes sense. But this town is also about perception, about how you make your case."

To date, MEDEA has released only two out of more than a dozen reports they have produced. One was "Scientific Utility of Navy Environmental Data," and the other was on the effects of Soviet dumping of chemical munitions in the Arctic environment. That report concluded that the risks to the Arctic seas are minimal. At the same time, the U.S. Navy and other government agencies refused to conduct a thoroughgoing study of impacts from nuclear and chemical dump sites off America's coastlines. These Cold War atomic relics contain tens of thousands of barrels of radioactive and hazardous waste from the navy and Atomic Energy Commission (later the Department of Energy). They are mostly scattered and unmapped in what are now national marine sanctuaries off northern California and Massachusetts.

Among the once classified military tools that are becoming available to science are the SOSUS (Sound Surveillance System) arrays, some navy ships, mapping technologies, navy sea-surface gravitational measurements (satellite altimetry), and sub-Arctic ice studies. The satellite measurements of the sea's surface are accurate enough to detect minute elevations of water that could reflect either submerged geological features like volcanic mountain chains, sea-level rise due to warming, or the bulk of a passing Typhoon class Russian submarine (not close enough to pinpoint but perhaps close enough to drop a nuclear depth charge in the area).

In 1998 and 1999 the navy also provided one of its nuclear submarines, the USS *Hawksbill,* for civilian scientists to travel under the Arctic ice and set up a polar surface camp. From the submarine they were able to study the thinning of Arctic ice associated with rapid climate change. On their second journey they surfaced at the North Pole for a memorial ceremony, scattering the ashes of Waldo Lyon into the frigid Arctic sea. Lyon was the civilian founder of the navy's Arctic Submarine Laboratory and a longtime advocate of polar exploration. Woods Hole's Al Vine got a similar and fitting sendoff back in 1995, when *Alvin* took its namesake's ashes down to the depths of the Mid-Atlantic Ridge, some 13,000 feet below the surface, where they were deposited along with a plaque in his honor.

With the decommissioning of the *Hawksbill* in 2000, the navy announced it could no longer provide nuclear subs for scientific research (except for NR-1, for which Bob Ballard seems to hold the pink slip).

"We have gone from as many as 96 submarines in 1990 to a projected 50 submarines by the year 2003," complained Rear Admiral Albert Konetzni, Jr., commander of the Pacific fleet sub force. "Planners are already being asked which valid missions the remaining submarines can fulfill and which missions will go unfulfilled."

"I guess I see evidence of global warming and think there may be something to this. But I don't see those issues as having any approximate effect on the navy or our national security decision making. We have too many snakes near the campfire to worry about those things," says Rear Admiral Andrew A. Granuzzo, the pugnacious New Yorker who is the navy's director of environmental protection.

This lack of interest in an issue that is expected to generate regional instability, humanitarian disasters, and thousands of storm-related casualties (as occurred in coastal Honduras in 1998, Venezuela in 1999, and Mozambique in 2000) is not limited to the navy. Despite a presidential directive stating that "environmental issues are significant factors in U.S. national security policy," less than 1 percent of the NRO's reconnaissance satellite time is dedicated to environmental data collection.

"The irony is environmental scientists are now the ones who have the global view," says Scripps director Charlie Kennel. "As the navy sees itself fighting wars in close waters, however, its interests should link up with our concerns. The navy will find itself with technologies that match our new environmental needs," he argues. "If the navy is sampling water, diagnosing it for signs of biological weapons, that kind of detection method can also be used for tracing pollutants or other scientific sampling. Resource management is also likely to promote conflicts in the future. And our job is not to turn our thoughts away from that, but to think it out. We will need real-time observation systems to cover environmental security issues like global climate agreements that have to be monitored for CO_2 reduction."

"The Cold War was the driver of oceanography. One of the new drivers may be found in population concerns," adds Bob Gagosian, director of Woods Hole. "If our population doubles to 12 billion and our coastal population triples in this century, it's not going to be enough to protect the oceans. We're going to have to learn how to manage and use them wisely which means understanding them far better than we do today."

Kennel, who was appointed Scripps director in 1999, and Woods Hole's Gagosian, appointed in 1994, represent a post–Cold War transition that is taking place in American oceanography. Kennel previously directed NASA's Mission to Planet Earth. Gagosian, an organic chemist, is interested in global change research and human impacts on the marine environment.

"Everyone says the future of oceanography is now in biology," explains Deborah Day, the Scripps archivist and institutional memory. "The future is in air-sea interactions and understanding climate change, and understanding the biological consequences of the breakdown in ecosystems and fisheries. How will we cope with this, and what are the consequences of the deaths of coral reefs and of cods?"

"What we're now doing is a bit like the exploration of the American West," Kennel tells me in an interview, with the blue waters of La Jolla cove glittering just beyond the picture windows of his waterside office. "One hundred and fifty years ago explorers caused massive biological reorganization of the continent [killed off the wildlife and chopped down the trees]. And behind these trappers came the scouts from the biological labs in England cataloging the biological wealth, studying it as they were destroying it. One hundred and fifty years later we don't know what's out there on the ocean frontier. Our scientists think most ecologies are collapsing. Look at the fish. On land, buffalo provided food for whole Indian cultures and now the buffalo are gone. The same's happening with our fisheries." He pauses long enough for me to think that perhaps this is his vision; oceanographers as obituary writers, recording the passing of a living sea. Then he rallies a bit. "Personally, I feel an ethical pull with these environmental problems. I find it very attractive to want to do something that might matter."

Still, much of the cutting-edge work in biological oceanography has shifted from the big marine science centers to smaller coastal programs and cooperative enterprises like MB-CORC, the Monterey Bay Crescent

Ocean Research Consortium, made up of 27 small- to medium-sized institutes linked by modem and mobility. Not surprisingly, two of the more
interesting if disturbing frontier discoveries of recent years, that of tiny
nutrient-enhanced predator microbes called pfiesteria, and, at the other
extreme, a 7700-square-mile nutrient-fed dead zone in the Gulf of Mexico,
were both identified by female scientists from less recognized institutes,
Dr. JoAnn Burkholder at North Carolina State and Dr. Nancy Rabalais at
Louisiana Universities Marine Consortium.

While the real action may now be in climate modeling, coastal processes,
ecology and marine biology, the old institutional players are not averse to
using the public's growing concern over the state of the Blue Frontier to
advocate for more big science projects in the deep oceans.

The idea of a fiber optic and satellite-linked, real-time ocean observation system spread around the EEZ and beyond has become a major goal
for academic scientists and administrators. Along with the directors of
Scripps and Woods Hole, Admiral James Watkins, the former chief of naval
operations and now head of the Consortium for Oceanographic Research
and Education (CORE), which represents some 60 university institutions,
is lobbying hard for it. Interest is also growing among the four Ns of ocean
science: the navy, NOAA, NASA, and the National Science Foundation.

While better understanding of ocean processes is always useful, the
question not being asked is how, without far stronger links between science
and public policy, can this ocean observation system function as anything
more than a high-tech version of those nineteenth-century biological scouts
who collected interesting samples on a dying frontier?

Despite a huge expansion of American oceanographic power during the
Cold War, when measured in terms of people, ships, and dollars, the public interest was not well served. The civilian oceanographic stations became
so much a part of the navy's blue-water world that they failed to identify
and alert the public to the Blue Frontier's nearshore living resources at risk:
the coral reefs, seagrass meadows, salt marshes, mangrove swamps, barrier
islands, beaches, watersheds, estuaries, and the once abundant marine
wildlife that depend on them. The responsibility for this may be limited but
we are all sharing in the consequences.

The Cold War may be over, but the navy continues to thrive in a world
of uncertainty and cost-plus contracting with, for example, an $83.5 billion
budget for the year 2000 (out of a $268 billion defense appropriation).
That is roughly 12 times the budget of the other two major ocean agencies,
the Coast Guard and NOAA, combined. Still, the navy, like almost all
bureaucracies, continues to insist that it needs more money to carry out its
mission.

I'm sitting inside a noisy narrow fuselage facing backward, feet up and arms crossed over restraining gear and life vest. Two crewmen in insect goggles and helmets wave their hands frantically to signal the time has come. There's a sudden yank and I'm watching my peripheral vision shrink as I feel the pull of 3 Gs, like being sucked over the falls backward on some 40-foot monster wave. Time slows down during the 2 seconds that the steam catapult accelerates us from zero to 150 miles an hour before tossing us off the lip of the USS *John C. Stennis*. But unlike an F-18 Hornet that would goose its afterburners and streak near vertical toward the heavens, streaming white contrails off its wingtips, our two prop COD (Carrier Onboard Delivery) aircraft seems to drift like a freed balloon, banking gently to port. A catapult-assisted takeoff is definitely an E-ticket ride, but unlike an earlier cable-arrested landing, none of my fellow passengers seems to have lost his lunch.

The last few days spent on the *Stennis* have been a fun and deafening experience, getting to see how a million dollars a day of our tax money is spent maintaining the smooth functioning of the $3.1 billion CVN-74. The *Stennis* is one of America's 12 aircraft carriers that, along with their $20 billion battle groups, ensure America's ability to project power anywhere in the world. The *Stennis* is named after a Mississippi senator who, as head of the Armed Services Committee, never saw a navy budget he didn't like, even as his state rated a consistent forty-ninth in the quality of its children's education.

I arrived on the *Stennis* in a mixed group made up of members of the Young Presidents Organization, top corporate types under the age of 50, and four guys from the Tailhook Association, the naval aviators group that gained infamy following its sexually predatory 1991 convention in Las Vegas. Fittingly, one of the outcomes of the Tailhook scandal (most of which involved a high-level cover-up) was the navy's accelerated program to qualify female combat pilots—women like Beth Creighton, a red-haired, self-confident flyer in a patch-covered jumpsuit I meet in the officer's mess. Flying the F-18 Superhornet FA 122 she has run a number of combat patrols over the no-fly zone of Iraq. Most of the pilots on the *Stennis* this week, however, are not veteran flyers but trainees doing their first at-sea landings and takeoffs or CQs (carrier qualifications).

The *Stennis* weighs 95,000 tons (about 20,000 tons shy of a supertanker full of oil), has a 4.5-acre flight deck, and can carry a crew of 5100. It is more than 24 stories tall, from the top of its mast to the bottom of its hull, including 18 decks of ladders. It is also about the length of the Empire State Building if you laid that building on its side and gave it the power to blow up the Chrysler Building, Rockefeller Center, and the UN.

Even more incredible, from my point of view, every year the weighty equivalent of 900 of these ships is taken out of the world's oceans by commercial fishing fleets. "And yet fishermen are still losing their jobs," the captain comments when I mention this to him. The captain, Richard K. Gallagher, is a lanky, narrow-faced man in his forties, with clipped mustache, receding brown hair, and a forthright manner that leaves little room for irony.

We're standing that evening on the darkened bridge of the *Stennis*, silhouetted in the red glow of night operations lighting, watching the periodic double white disks of F-18 afterburners as Hornets are catapulted off the deck below us, their onboard fly-by-wire computers controlling their initial seconds of flight. A former commander at the Top Gun school at Miramar Naval Air Station, Gallagher (flight name, "Weasel"), also commanded the antimine USS *Inchon* in the Adriatic during the war in Yugoslavia.

The *Inchon*, a converted helicopter assault ship, has 10 large mine-hunting CH-53 helicopters, two Seahawk spotter-rescue choppers, and four explosive ordnance dive teams, each made up of 32 divers with fast rib boats and mobile (flyaway) decompression dive lockers. It also carries ROVs and occasionally works with the navy's marine mammals (in the near future the navy may forward deploy its dolphins and sea lions on ships like this).

"Each CH-53 helo could lift 30 to 40 tons," Gallagher tells me. "So we flew food from Tirana Airport in Albania, moved food into refugee areas. Remember we feared hundreds of thousands of people might starve to death. It was a war tactic of Milosevic to drive these people across the border and there was lots of worry about people starving and a big concern to get refugee camps going before winter."

"Which is now all part of littoral warfare?" I ask.

"Well, we've been operating in the [Persian] Gulf for some time," Gallagher replies. "I was ExO [executive officer] on the [aircraft carrier] USS *Eisenhower*, operating in the Red Sea when Iraq invaded Kuwait. We've been training for years to operate in these more restricted waters. It hasn't taken us by surprise."

In the near future the navy plans to fight more coastal engagements with Virginia-class, shallow-water submarines equipped with cruise missiles and contingents of SEAL teams and AUVs (autonomous untethered robots), and with 35-knot-plus Cyclone-class coastal patrol ships, missile-firing arsenal ships that bear a striking resemblance to giant Civil War ironclads, electric-drive destroyers, and new marine amphibious ships carrying tilt-rotor Osprey airplanes, air cushion landing craft, and better-armed and armored amphibious assault vehicles. There will also be a new generation of carriers for air support (the General Accounting Office does not believe they need to be nuclear powered but the navy insists they should be). The fighting will be coordinated through real-time battlespace intelligence generated by sub-launched robot aircraft and satellite imaging in 3-D high-definition TV. This will allow all the ships and planes, in the words of former chief of naval operations Admiral Jay Johnson, to "distribute firepower in new and exciting ways."

The only thing that will not change is people getting blown up and dying in war, although sensors woven into the body armor and underwear of young men and women could tell medics where they have been hit and how quickly they are bleeding out. If more than 20 percent of the casualties of these future conflicts involve combatants rather than unarmed civilians, that will also mark a dramatic change from late twentieth-century warfare.

"There is a mindset change taking place," Captain Gallagher tells me. "Now we're not sure who's interesting, who's the next threat, where's the next hot spot, and how to prepare, because there are some good weapons systems proliferating out there."

"We sell a lot of them," I point out.

"We've shown we can operate in the Gulf, in the Red Sea. When you don't have a monolithic threat, however, it makes it difficult for people to remember we still have skilled adversaries."

"None of whom have aircraft carriers."

"Well, the Russians have one, though I'm not sure if it's operational or could be made so. But the French also have one that's supposed to come on line, although people have been waiting on that for years."

"How about the British?" I offer helpfully. "We've fought them twice."

I'm not sure if it's something I said, but my stateroom that evening is changed to directly below the night-launch catapult. Unable to write or sleep to the whoosh-bang of landing jets, the roar of afterburners warming blast deflectors, or the steam scream and clang-clang subway-like return of the cat cable, I decide to head up to the air control tower.

The highest enclosed space on the ship's island (the superstructure), this roost is the domain of air boss Mike Allen and miniboss Baron Asher. Close to a dozen of their people also crowd around. I look out to see an F-18 miss all four arresting wires, its tailhook scraping the deck, sending up a shower of sparks before the pilot uses his afterburners, pulsing like a pair of white hot dragon tails, to drive the 25-ton plane back off the deck. "That's called a bolter," Baron explains to me. "Really chews up the deck. We have to continually resurface."

The next aircraft, an A-6 Prowler makes a good landing, getting yanked to a stop by one of the arresting wires. The deck crew quickly push the plane to the black water edge of the flight deck to turn it around, as its wings begin to fold upward. "We've got limited acreage to work with," Mike Allen explains. "The way the planes land is they watch a set of lights off the side of the carrier. The center light is called the meatball. It gives the pilot a target, a 13-by-13-foot box from his eye to the deck. If he goes in too low, he hits the fantail; too high, he misses the wire; right or left, he hits other aircraft. In five days of these CQs we'll do 700 of these landings."

"How many launches?" I ask, as I scribble furiously in my notebook.

"We try to make our launches and landings even," he grins hugely.

Mike is big and jut-jawed, clean-shaven with a rough dinosaur look about him, a 20-year career man. Baron is shorter and slighter with a light beard, glasses, and 18 years in the service. They can handle as many as 47 aircraft and 200 crew moving on the deck at any time. There are 15 planes being launched and recovered tonight.

"Attitude," crackles the radio. It's a signal officer talking to an SA-3 pilot, who adjusts his wings and makes a rough landing. His landing space is 800 feet long.

An SA-3 bolts it. An EA 6 Prowler bolts it. An F-18 Hornet is shot off the catapult, afterburners flaring.

"How much fuel do planes like that burn?" I wonder. "An F-14 Tomcat can burn 300 gallons a minute with its afterburner. It can burn its fuel in 8 minutes or 2 hours, pilot's choice," Mike says. "When our air wing's operational, we burn 200,000 gallons a day. We have about 20 million pounds of jet fuel on board."

Another jet rockets off the deck.

"What are the worst sea conditions you'll launch in?" I ask. "We'll fly with a 200-feet ceiling and half-mile visibility during operations," Mike replies. "We've launched with water coming over the catwalks of the deck [60 feet up]. We wait on the swell till the bow is through the trough and shoot on the upswing. We've done it here and in the South China Sea on the backside of a typhoon."

The next afternoon I'm back in the tower. It's jammed with reps from Shamrock Squadron 41 out of San Diego's North Island Air Station trying to get Mike to qualify their people. "331 Boss—four total, he needs two more traps, sir." "Wave off 306 on deck time. 225's gone off." The planes are coming every minute or so now. "242—Billy's got two so far, he needs two more, sir."

"565 is leaking fuel on the deck," a radio voice announces.

From the tower you can see the roll of the ship over the bow. "Name is Breckenridge, sir. He needs two and three at a minimum." "336 bolted." Another plane lands. "306 on deck, time one zero." "227 bolter, bolter."

"342 has an external tank problem," a radio voice reports on the fuel spill. A plane comes in low and is waved off.

Mike looks up at a jet flying high overhead. "What's that guy doing way up there, gol darn it."

Planes are coming in every 45 seconds now.

"Check it out," Baron says glassing the horizon with his binoculars. All I can see is one of two Seahawk search-and-rescue helicopters waiting on station, and then I spot the vapor blows and curving backs of gray whales. At 20 knots we quickly pass the pod.

"We were seeing a lot of whales earlier this summer," he tells me.

Later, I ask Lieutenant Nick Fiore, one of the chopper pilots, if he enjoys spotting marine life. "Well, you're circling out there for hours," he says. "If you see whales or dolphins that's fun for like fifteen minutes. Then you want to take a gun to your head. You're always circling."

I'm back in the tower that evening. I figure this is where the action is. Mike and Baron haven't left, of course. Number 242 is coming in for a landing. Baron explains the psychology of a night landing. "Once you get past the idea that you're hurling yourself at the water at 500 feet a second, it's not so hard to land."

Mike is staring at the approaching plane. "Hummm, humm, humm." Baron laughs. "Boss is sending out good vibes." Number 242 makes the trap and is quickly shoved aside by the deck crew. The reflective tape on their float coats make them look like fireflies as they push the planes around. Eight planes are now lined up behind the two pop-up jet blast deflector walls, waiting their turn to launch.

"Strobe back lights, turn them up a tad. Okay, that looks good," Mike instructs, as a narrow strip of landing lights brightens infinitesimally. As 227, an F-18 Hornet, is launched, one of its engines blows out with a bang—then, a second and a half later, relights.

There's silence in the tower, followed by a noticeable break in the tension. "It's an air burp from hesitation in the fuel air mix, maybe a feedback from the catapult steam," Baron explains. "It's hard to distinguish it from an engine failure is why the concern."

Even with the obvious professionalism of carrier ops, this is still a high-risk enterprise. Since its first deployment in 1998, the *Stennis* has lost an S-3 pilot and navigator, killed when their plane dropped off the bow, and two deck crew mutilated, one losing part of a leg, the other an arm that was surgically reattached, after a blast deflector crashed down on them.

The next plane coming in drags and sparks its tailhook before bolting back off the deck. "725 bolter," says a young sailor with a clipboard.

"A new voice heard in the tower," Baron smiles at the sailor who beams back like his dad just bought him a new baseball glove.

"336, 727, 362, touch and go, 53, touch and go," Mike instructs. They're taking them around a race course pattern, giving them a new play.

They wave off a plane coming in too low that had bolted a couple of times during the afternoon. "Mr. Breckenridge storms onto the night stage," Baron notes wryly.

You can't help but like guys like this, the air boss and miniboss working well together with a kind of casual discipline, controlling both ends of the deck simultaneously, fully confident of each other's calls, moving a deafening carousel of prancing, bolting, jolting high-performance jets like so many overpriced Lipizzan horses going through their paces.

Back down below deck I find two pilots talking in the passageway outside my stateroom, too pumped to sleep. Dave Bigg and Bryant Medeiros are both 29. Bigg made his first carrier landing on his first try, then bolted twice tonight. "You come down at 150 miles per hour and see a little carrier box in a big black ocean, and you're looking into a geometric cone [he creates one with his hands], and you hook it to your eye, and by the time your eye is over the ramp it's a 3-foot cone, and every foot your eye is off puts you 15 feet further down the deck."

"I did a touch and go, two bolters and a trip," Medeiros, an S-3 pilot, volunteers. "I bounced over the four wires on the touch and go. That scared me." Bigg tells him of an acquaintance of theirs whose wing blew up on a recent landing with the landing gear shooting through it. "But he climbed out okay and may get to fly again after the incident investigation."

"Night landings are not fun. It's the hardest thing you can do," Medeiros says.

"But you did it."

"Honestly, you forget the carrier is there," Bigg jumps in. "You're concentrating so hard. You're looking at the stupid ball and white line [flight deck center line] between your legs. That's it." He shakes his head. Medeiros grins. I retire, leaving them to talk out their adrenaline jag.

The next morning I go up on deck with a new group of VIPs— Republican congressional aides from Washington. We move single file behind our guide. Even with our Mickey Mouse ear protectors, the place is roaring with A-6s, SA-3s and Hornets slamming the deck and all sorts of catapult steam, loud noise, and unburned fuel odors hitting us with the rush of jet exhaust and young sailors directing planes and giving each other high fives. Everyone gets to take some snapshots of the launches and color-coded crews. Brown jerseys are plane captains, who prep the planes. Blue vests place chocks and chains; yellow are directors (or shooters), the ones who do the crouch and point for the launches; green are maintenance people; purple, fuel handlers; white, safety and medical; red (and silver foil), the crash, fire, and salvage crews, hanging out on their crash carts.

The average age on deck is 19, and the kids seem to love it, young guys with goofy short haircuts and pretty girls draped in tie-down chains and greasy cammo pants. In between the launches and traps, we are moved back to the arresting wires (ducking under a jet's tail on the way). An F-18 turns and we're all waved into a crouch as its powerful exhaust whips over us. Seeing half a dozen congressional aides forced to their knees makes the whole trip seem worthwhile.

Although not allowed to see the nuclear reactors, encryption room, or weapons magazine (which we're told is "inconvenient"), we still get a pretty good Cook's tour: visiting the Combat Direction Center, air ops room, a pilots ready room, the hangar deck (where a step-aerobics class is underway), overcrowded crew quarters, jet engine shop, cryogenics plant, and rudder stems (the rudders are 60 tons each). I ask to visit METOC (the meteorological and oceanographic office), where I talk with Senior Chief Glen Picklesimer about tactical exploitation of the environment.

Despite the navy's major impact on civilian oceanography, the oceanographer of the navy has historically been an undervalued command. Until recently, the position was seen as a dead-end job for two-star admirals not expected to make their third star. The low point came in the wake of the Tailhook scandal, when the navy inspector general, under fire from Congress, was given the job as a sinecure until his retirement.

But the new tougher requirements of coastal warfare have begun to turn things around. "It's a lot more challenging getting into the littoral," Picklesimer explains. "We used to chase subs in blue water with 10- to 15,000-feet depths. Now we get up near shore, and it's like being in mountain ranges with biologics [animals], waves, bottom sound, ice, all these interesting noises. All our oceanographic ships are spending their time doing surveys in the littoral these days."

At lunch the assistant propulsion engineer tells me of pumping radioactive waste water into the sea. "But at no more than background levels, no different than this," he assures me, holding up a glass of water with a lemon wedge and ice.

This reminds me that I want to talk to the ship's environmental officer. Lieutenant Commander Paul Kratochwill, who takes care of all nonnuclear engineering equipment, is as close to an environmental specialist as the ship has.

"When I joined [in 1985] we threw anything and everything over the side 12 miles out," he tells me. "Then, in January 1997 you had the international convention on dumping go into effect [which bans the dumping of plastic at sea]. All our food products come in plastic. We serve 18,000 meals a day. So, what we do now is rinse and wash the plastic and use these 400-degree ovens to make plastic pucks." He takes me to see the ovens and piles of giant pizza-sized pucks. "We produce around 150 pounds a day," he tells me. "We had a ship that dumped some medical waste overboard a few years ago and that had bad publicity value for the navy, so things have begun to change. With paper we mulch it and turn it into a slurry that goes over the side. Cardboard is burned in this incinerator." He shows me the sealed room next door where a sailor in a balaclava (hood) is using a heavy metal rod to load the blazing incinerator. "Join the navy and see the world?" I ask. "It's not a nice job. That's why we only make them do it three months at a time."

"Burlap bags of glass and metal are dumped over the side," he continues. "Thousands of arresting wires [greasy inch-and-a-half-thick cables] litter all the sea bottoms. We get rid of them after a hundred recoveries [of aircraft]. Oily wastes we dump out beyond 50 miles."

The sewage for the *Stennis*'s 5000 people is mulched but not treated before being pumped over the side, 3 miles or more offshore. Cruise ships, by contrast, which are in almost constant operation and might carry 3000 to 5000 passengers and crew, are equipped with secondary sewage treatment facilities.

"I'd say 30- to 40,000 gallons a day would be a low side estimate," Kratochwill calculates the *Stennis*'s "black water" sewage production. "In U.S. and Canadian ports we pump into sewers or sewage barges. In the Philippines and other places we used to just pump it over the side. Japan now has some new laws that won't allow direct discharges like that. And I remember we also had to hire sewage barges in Hobart [Australia] last year."

"I don't think we've dumped raw sewage over the side in a port in over 20 years," says Rear Admiral Andrew Granuzzo when I later ask him. Granuzzo oversees the navy's environmental and occupational safety programs from his offices in Crystal City, Virginia. "The fact is we're going overseas and investing in infrastructure improvements [of sewer systems] for ports like Naples, Cannes, Greece."

"What about the aircraft carriers themselves?" I wonder.

"Have you seen the sewage plants on those cruise ships? They're big and labor intensive things. We've tried sewage treatment plants. I got one on my amphibious ship [one he used to command]. It went well but they're highly technical and take a lot of care and feeding for our big ships. The tradeoff in size and weight kept us from going after them.

"We are investing in new incineration technology instead. We hope to have self-contained membrane technology to revolutionize waste disposal and think it can also work for bilge and oily water waste and also appears to be applicable to black and gray [sewage and shower] water. So then we can incinerate it all."

"We're developing the environmentally sound ship for the twenty-first century. In the next 15 years we'll build ships with zero discharges," he claims.

"We used to dump stuff without thinking, but times change," Lieutenant Commander Kratochwill muses. "When I was an ensign in 1986 I took a boat over to the Thirty-second Street Naval Station [in San Diego], and there was nasty dirty stuff in the water, oil, and solids, and today you can see the bottom there, no garbage or oil at all."

Still, because the navy was exempt from most environmental laws during the Cold War and operated with little concern for the impacts of toxic paints, solvents, antifouling agents, munitions, or low-level radiation, bottom sediments in military ports and shipyards are among the most toxic in America. A 1998 NOAA study of sediment toxicity in 22 harbors found San Diego to be the second most polluted, after Newark, New Jersey, with major hot spots located in the sediments around North Island Naval Air Station and the Thirty-second Street Naval Station.

In December 1997 port dredging for the arrival of the *Stennis*, the first of three Nimitz-class nuclear carriers to be permanently stationed in San Diego, was halted when the sand being used to replenish an eroding beach in the north county town of Oceanside was found (by curious children) to contain unexploded munitions, including 20mm and 50mm rounds.

The navy's use of offshore islands for bombing and live-fire exercises has also generated controversy from Kahoolawe island in Hawaii (which was returned to the state and is now being cleared of ordnance) to Vieques island off Puerto Rico, which became the focus of widespread protests in 1999–2000 following the accidental killing of Vieques resident and security guard David Sanes by a navy bomb.

"Vieques is an environmental success story. It's the jewel of the Caribbean, because of navy stewardship," insists Admiral Granuzzo. "If you compare and contrast it with its sister island we had to give up, Culebra [a major tourist destination] has become an environmental disaster area. On Vieques we're protecting the sea turtles. It's a beautiful tropical island. Only the very western tip is a target range." Still, two-thirds of the island is controlled by the navy, with its 9400 civilian residents confined to the center of the island.

Along with its protected beauty, Vieques has an "ideal moderate tropical climate [that] permits year-round ops with practically no cancellations," or so the Navy used to advertise to foreign military services wanting to rent the site for their own live-fire war games. For those foreign fleets who came, the navy promised "excellence in all warfare areas."

I ask Lieutenant Commander Kratochwill about the *Stennis*'s airborne emissions from burning 200,000 gallons of jet fuel a day off the coast of southern California. "No one ever asked me about that before," he admits.

In negotiating the Kyoto Treaty on greenhouse gases, the United States insisted that military bunker fuels and strategic (nonshore-based) forces be exempted from any reductions. The navy is also exempt from the 1990 Oil Spill Prevention Act, Marine Pollution Act, Aquatic Nuisance Act, and Superfund legislation.

Still, the navy's internal handling of fuel and oil spills (as opposed to sewage) has won praise from the Coast Guard, NOAA, and other observers. Their strict reporting rules on discharges stands in marked contrast to Royal Caribbean and other foreign-flagged cruise ship operators that have faced multimillion-dollar fines for illegally dumping oil and hazardous chemicals in U.S. waters.

I go back on deck to watch the U.S. navy ship *Pecos* extend its hoses across the blue swells and begin pumping 1.5 million gallons of jet fuel to the *Stennis*. From across the flight deck I can see the 400-foot supply ship's bridge rising and falling with the 10-foot swells, which are hardly detectable from the surface of the giant carrier. Looking back over the stern, I watch our wake leave a wide bright aquamarine trail in the ocean.

I walk over to talk to a young maintenance crew gathered around a plane, including avionics specialist Laurie Hale. She is proud of the EAB6 radar-jamming planes she works on. "They can show a bunch of our planes on an enemy radar screen, even if there's really just one. They went out with all the flights in that last little deal."

"Last little deal?"

"You know."

"You mean the war in Yugoslavia?"

"Yeah, that deal."

I wander over to where the *Pecos* has just disconnected its fuel lines.

"It's kind of sad," says John Barbour, a member of the Young Presidents group, a thick-set Scotsman, and owner of a toy company who with his camera gear and shooter's vest looks more like a war photographer than many a war photographer I've known.

"What's sad?" I ask.

"The redundancy and the waste. It's impressive, but you think of the poverty and the shitty education kids get in this country and this all seems too much."

Just then, as the *Pecos* drops back, the *Stennis*'s sound system starts playing U-2. "I have run, I have crawled, I have scaled these city walls, only to be with you." With the oiler slipping behind in the steely blue sea it is a strangely moving moment. "But I still haven't found what I'm looking for. But I still haven't found what I'm looking for."

The ocean rolls, the smell of jet fuel briefly lingers in the air. Those hydrocarbons are the combustible sinews of industrial civilization. Modern war and the power to make it depends more than anything else on that single product, a product that, like the U.S. Navy itself, continues to transform what is happening on, above, and below America's Blue Frontier.

[Chapter Five] Oil and Water

Let's remember we fought a war a few years ago over one thing and that was oil.

—SENATOR FRANK MURKOWSKI OF ALASKA

Then this Hollywood star pulls up in his limo, must have been half a block long, wanting to know what we've done to his beach. And I'm thinking, hey that limo of yours doesn't run on sunbeams you know."

—OIL COMPANY SPOKESMAN RECALLING THE
HUNTINGTON BEACH OIL SPILL OF 1990

I'm staring into the 400,000-gallon main tank at the New Orleans Aquarium. It's swarming with tarpon, redfish, snapper, stingrays, alligator gar, sea turtles, sawfish, nurse sharks, black drum, and horse eye jacks, all swimming through the re-created metal legs of an offshore oil platform. The sign off to the side reads, "Gulf of Mexico sponsored by Amoco, Shell, Exxon, WMP, Chevron, Kerr-McGee and Tenneco."

Flying out of Venice, Louisiana, the next morning, I'm not sure if that sign referred to the exhibit or the real thing. From our Bell 412 four-rotor helicopter, a kind of buffed-up version of a Huey, we pass over shredding islands of brown spartina or salt grass, cross-hatched with canals and studded with oil tank transfer stations. South Louisiana is losing 30 square miles of land a year, about a football field every fifteen minutes, due to sediment loss from levees, oil and navigation canals, subsidence, and sea-level rise from the burning of fossil fuels. There is a proposal to invest $14 billion to try to restore the Mississippi Delta and save the land, but with 80 percent of the waterfront owned by global oil companies, there is less national will to save coastal Louisiana than the Everglades National Park of Florida. A flock of large white egrets flies over water beneath us, like stately extras from some *National Geographic* documentary.

We pass over the southwest channel of the Mississippi and a breaking surfline the color of chocolate mousse. As we fly beyond our first cluster of platforms the water turns a strange jade green. Soon we're some 50 miles offshore in deep blue water. Looking out across the horizon you notice there's no point at which you can't see oil platforms. There are some 4000 platforms operating in the Gulf of Mexico today. Offshore drilling accounts for 20 percent of U.S. oil production and 27 percent of its natural gas. Despite heated debate over drilling off California, Florida, Alaska, and North Carolina, 93 percent of all present offshore production takes place in the gulf. In the early 1990s there were some reports that the gulf might be tapped out after a generation of exploitation, but that was before the boom in deep-water drilling (1000-foot depths and beyond).

We circle around a flat-topped platform called Pompano. Owned by BP-Amoco, it is the second tallest bottom-fixed structure in the world, drilling into the ocean floor 1310 feet below the surface. About 700 feet wide at its base, it is taller than the Empire State Building (or the *Stennis* standing on its tail). With newer tension leg platforms (TLCs), giant water-filled spars, and proposals to bring production ships called deep draft caisson vessels (DDCVs) into the gulf, drilling is rapidly moving toward 10,000-foot water depths. In the fall of 1998 the old CIA spy ship *Glomar Explorer*, leased from the navy by Global Marine and converted into a drilling ship, sank a well in 7718 feet of water. It then drilled 10,000 feet into the sub-seabed rock in an unsuccessful attempt to strike oil and gas. Still the prediction is that deep-water drilling will produce 15 to 25 billion barrels of gulf oil in the next 20 years, about as much as has already been taken out since the industry went offshore back in 1947.

We land on Pompano's helideck 12 stories above the water. Even with the copter's rotors stopped the sea winds continue to whip against us at 30 knots. We climb down two levels past some rigid enclosed lifeboats to the living quarters, walking on cookie-cutter grating that lets you see all the way down to the swells breaking against the platform's legs. Entering the crew structure, we pass a three-button panel marked "Abandon Platform, Fire, and General Quarters." The TV room–galley, with its cafeteria-style service, metal tables, bug juice dispenser, video player, and thick couches grouped around the oversized TV, reminds me of a number of work boats I've been on, minus the sense of movement.

"It's just like an aircraft carrier in that the platform has to be completely self-sufficient," Hugh Depland, BP's public relations guy, tells me.

He then goes off with a three-man video crew that is shooting a company-sponsored spot for the Chicago Museum of Science and Industry. Oddly, all three of them are named Mark.

So is one of the two helicopter pilots who flew us out. Mark Stearns is a short, bright-eyed oil-patch veteran with a snow-white mustache and 16 years of experience in the gulf. Cathal Oakes, his lanky, younger Irish colleague, is studying up for his final pilot's exam. They both live in California and work in the gulf, 14 days on and 14 days off. Most rig crews work seven and seven.

"There are probably 150 to 200 aircraft in the air at any one time out here, moving people on and off the platforms," Mark tells me. "And the DEA [Drug Enforcement Agency] watch us all, track us by transponder. If you go south of a certain point in the gulf or if your tracker is turned off for any reason, your company gets a call, right away."

When not flying for ERA, one of the petroleum helicopter services, Mark does water bucket drops for the U.S. Forest Service. He fought the big Florida fires of 1997–1998, caused by the El Niño drought. "Watching rainforest burn is pretty amazing," he says. "Those palmettos really burned hot."

I ask him if helicopters are less safe than airplanes, as I have always believed, and he reassures me that hour per air hour what they do is as safe as flying a commercial jet. This despite the fact that I was advised not to wear my steel-toed boots on the flight out (they weigh you down in the water), and just about everyone out here seems to have a crash story.

Preston Smith and Shelby Williams were both on a Bell 407 that crashed on September 18, 1997. "All five of us survived," Preston tells me. He is a thick-set man, probably in his late fifties, bald, with silver glasses, a flushed face, and nervous hand tremor. "We'd just left 826 platform when we hit the water. . . . We all got hurt. I had bruising in the chest around my heart and my back was hurt, my knees, and I still have numbness in my lower back. It's also caused me some mood swings and stuff I'm still not over. I'm on light duty out here."

Was he willing to ride copters after that?

"No, but this is the only kind of work I know how to do, so I have to support my family. I just have to face forward and sit by the door is the only way I can ride them now."

"We were 10 minutes into the flight and just heard a loud bang. The tail rotor had broken off and sliced into the boom," recalls Shelby Williams, a handsome, broad-cheeked, brown-eyed African American in his thirties, with closely cropped hair under a Bud Lite bill cap. "We angled over steeply, breaking left at 140 miles per hour and then started looping down. We went in nose forward and hit the water at about a 75-degree angle. But the pilot hung on to that joystick even seeing this wall of water coming at him. He was really good.

"No one ever yelled as we went down, we just followed our training, which says don't panic," he continues. "When we hit, the blades were still turning. There was a big buzzing from where the radio went out and then we went under the water, and the pilot deployed the floats and we popped up again. We threw the life raft into the water, pulled it back by this long lanyard, and stepped into it and just floated away. About 15 or 20 minutes later a copter spotted us, then two more came, and later there was a work boat. It took us to the platform and then another chopper took us to Jefferson Medical Hospital. I'd been flying backward and took the crash force from behind. My ligaments took the impact. It felt like someone was shoving needles into my back." Having since made his peace with helicopter air logistics, Shelby goes on to tell me how the next few years "are really going to be happening here in the gulf," thanks to the industry's new deep-water discoveries.

Normally operated by a crew of 12, Pompano is crowded with 22 extra men this month reconfiguring the platform for the return visit of a drilling rig. After five years of operation, its production has declined from around 68,000 barrels of oil a day to about 46,000 barrels (and 63 million cubic feet of gas). Not bad, at more than a million dollars worth of product every 24 hours, but it can do better, and will. The oil companies are now able to find oil- and gas-laden sands that they used to miss using 3-D seismic imaging and computer-controlled sensors. For older platforms like Pompano, they use what they call 4-D seismic studies, incorporating past production patterns into computer analysis of where additional hydrocarbons might be found.

Down in the highly automated Multi-Control Center (MCC), I meet George Yount, the operations supervisor. He's wearing a tan Cardhardt work coat and BP hard hat and looks like a beach master elephant seal, thick-throated, well-padded but strong, with a scraggly mustache and three-day growth of beard. He's been 25 years in the industry, starting as a rig deck roustabout.

Also working here is Wendy Lemoine, a thin, blonde assistant engineer. While the oil patch has been racially integrated for some time, it is well behind the navy when it comes to women. Wendy is the only female among some 80 men on the two platforms I visit, a not untypical ratio. A chemical engineer on temporary duty, Lemoine says that while she doesn't mind the work, she is definitely looking forward to getting back to her base in Houston.

After making sure I have a hard hat and earplugs, George takes me down to the well bay to see the Christmas trees (well pipes). On the way I look over the side and spot about 200 good-sized fish schooling around one of the yellow platform legs. A little further out the torpedo-shaped bodies and yellow tails of a pair of dolphin fish (mahimahi) streak by. Later in the day we spot a big manta ray cruising the area, its 9-foot wingtips clearing the water like sails. While platforms have not been shown to increase fish productivity, they do tend to concentrate fish, as do any structures in the ocean. The growing surfaces they provide for algae and barnacles attract small fish, which attract larger fish, which in turn attract recreational fishermen, who have become major advocates for the rigs.

Back in the 1980s some oil platform crews would reel in large amounts of redfish, snapper, and other commercially valuable species, loading hundreds of pounds of filets into big ice chests and taking them back to shore on crew boats where they'd sell them illegally to commercial fish houses. But a crackdown by the feds and the companies (after complaints from commercial fishermen) put an end to the practice.

The Christmas trees on Pompano are 23 vertical well pipes (plus two water reinjection pipes) married to small chokes and connectors so the oil can be separated through heater-treater processors and the gas dewatered before being pumped into big 12-inch pipes running to "the beach."

George turns on a small caffe latte–type spigot to show me the raw crude, a light-colored mix of oil, water, and gas that he lets run over his fingers. BP, which used to dump its processed water over the side, now reinjects it into the wells to keep the head pressure up. In the 24 hours before I arrived, Pompano produced 46,641 barrels of oil, 63,887,000 cubic feet of gas, and 15,692 barrels of subseabed water.

Along with wells drilled from the platform, Pompano has a tieback pipeline to eight subsea oil wells in 1850 feet of water four and a half miles away that were drilled and installed by ship. A new platform under construction will have a 30-mile subsea tieback. Having seen the drill deck (living quarters) and production deck, which also houses the electric generators (the platform operates on 3.2 megawatts of power), George takes me down to the subcellar, where the fire pumps, hydraulics, and utility equipment are located. There I get to check out the large gray pipes that drop to the seabed before running the oil and gas ashore. The bottom of the gulf is spider-webbed with tens of thousands of miles of pipes like these, along with underwater well heads and production complexes.

"A platform like Pompano costs around $350 million to build and operate," Hugh Depland tells me. On the horizon we can see Chevron's $750 million Genesis Spar (supported by water ballast and mooring lines), operating in 2600 feet of water.

Mark Moore, the video cameraman, tells me how while shooting an industrial video about the construction of Texaco's half-billion-dollar

Petronius platform, its derrick barge accidentally dropped the crew quarters, worth $70 million, that quickly sank out of sight.

That evening we eat a tasty dinner of jambalaya, crawfish étouffée, corn bread, french fries, and ice cream. Rig dining may not be heart healthy but at least none of the guys out there appears to be undernourished. I ask George about accidents on the rig, and he embarrasses one of the kitchen crew by recounting how the fellow injured himself zipping up his pants, and how the catering company required George to write up an accident report. Not done having fun, George turns to Hugh and says, "In the book he's working on," pointing at me, "he's writing about the navy. So I told him how he can make the connection to this place. Write about the navy, then the Coast Guard, then oil spills, then us." Hugh the PR man smiles wanly.

We go to sleep in double-stacked steel shipping containers converted to crew quarters (bunks, plumbing, and a washer-dryer). The next day we take the 15-mile helio-hop from Pompano over to Amberjack. I learn that rigs are named after their lease sales, which, for security reasons, are given theme-based designations by the oil companies' secretive exploration departments. That way if some drunk is overheard in a Houston bar mentioning how many millions his company bid on Bullwinkle or Nirvana, it won't mean anything to the eavesdropper. Lease sale themes have included rock bands, country-western singers, types of cows, booze, game fish, and cartoon characters. To date, none has been named after a famous woman author or environmental hero.

Amberjack is the ultimate Tinkertoy. An active drilling rig, it towers 272 feet from the waterline to the top of its bottle-shaped derrick. Its density of utilized space is a structural salute to human ingenuity. The rig has a four-story metal crew building, helipad, flare-off tower, tanks, processors, compressors, a drill deck with 8300 feet of piping stacked 12 feet high, 1000 barrels of drilling mud, mud shakers, cement, two big yellow cranes, an office shack, lifeboats, and hundreds of other flow-pipes, tubes, racks, gears, lines, and computerized systems hanging out over either end of its legs on wide, thick steel shelves. You know that whoever designed this thing does not waste closet space at home. Still, from the air, Amberjack looks small and somewhat fragile set against the vast, white-capped expanse of the gulf's deep blue waters.

The winds are howling close to 40 knots today, the swells are about 12 feet, and, with an extra half million pounds of drilling gear on board, you can feel some sea movement on this platform. Once inside we are given a safety lecture and told to remove rings and Velcro watch bands to avoid "degloving injuries," where the skin and muscle can be ripped off your hands. We're introduced to Cary "Call me Buba" Kerlin, a red-faced, spherically shaped "company man."

"Might look a little dirty," he warns us. "We've been getting a lot of gumbo mud while we've been drilling." Gumbo is a heavy, clay-thick, gray-black mud that is hard to wash off.

As a drilling supervisor, or company man, Buba has been around the oil patch, having worked in Colombia, California, and Alaska, as well as in the gulf. Before that he spent 12 years with the U.S. Fish and Wildlife Service doing environmental assessments on oil company dredging canals, "til they made me an offer to get out of government and into industry and I became oil trash," he grins.

Under the company man is the tool pusher, or rig manager. Then there's the driller who controls the drilling console, the skilled roughnecks who work for him, and the less-skilled roustabouts or general assignment workers. There's the mud man, or fluids engineer, who runs the lubricating muds (polymers, clays, dirt, and additives) that circulate down the pipe string. Several stories above them all stands the derrick man on his monkey board, a small catwalk from which he handles the high end of 42-foot sections of pipe. As the pipe tilts up toward him, he leans out almost horizontal in his harness to grab the top of the pipe and align it with the heavy rubber fill-up tool that adds drilling mud to the pipe string.

Buba takes me up to the drill deck. It's a noisy, thrilling scene; a choreographed dance of steel pipe, muscle, and machine. The cranes lift the pipe to the roughnecks and roustabouts in their hard hats and steel-toed boots who manhandle it into position below the derrick with its massive yellow top drive and block. I stand near the console on the water-slick deck watching the crew work the hydraulic tongs around the pipe stem and thread it into the hole with a creaky slow rotation before the top drive begins its work. At this point they are down to 8387 feet. With more than 36 other wells down there, it's a directional driller's nightmare. This hole is being drilled at a 45-degree angle, although they are capable of slant drilling like a boomerang, going down and then up again. One of the crew has a T-shirt that reads, "New Rig, New People, New Records."

Another 42-foot section of pipe is chain-winched onto the derrick floor like a skidder-pulled log coming up a clear-cut hillside. I move forward and begin taking pictures of the red-helmeted derrick man as he leans out from his monkey board like a trapeze artist to grab the 13-3/8-inch pipe top and begins shifting it around to line it up with the rubber mud hose dropping down on him from above. I'm carefully lining up my shot when one of the roughnecks sneaks up behind me and slaps my ribs, letting out an animal howl.

I turn around quizzically. He's grinning happily. "I can't believe you did that," another guy semi-shouts to be heard. I didn't jump at the prank because I knew there were no howling predator animals lurking on this rig—other than these guys, of course. On the way back to the helicopter I spot the crane operator on a break, standing on the catwalk outside his cab, licking an ice cream cone, and staring off into the Blue Frontier.

For more than half a century the leasing of offshore oil and gas has been one of the linchpins of U.S. oceans policy. In 1896, less than 40 years after the first rock oil derrick was drilled in Titusville, Pennsylvania, the first off-shore drilling piers were built out from the newly established spiritualist center of Summerland, California.

Soon the more secular oil men were at war with each other, hiring armed thugs, sabotaging each other's piers, and racing to suck up as much oil as possible before the wells lost pressure and had to be abandoned. Abandoned wells and badly managed gushers soon led to widespread oil pollution and fouled beaches. "The whole face of the townsite is aslime with oil leakages," reported the *San Jose Mercury News* in 1901. The resort town of Santa Barbara just up the coast quickly moved to ban oil piers, fear-ing their impact on tourism and beach life. By the 1920s the state of California had made several feeble attempts at regulating offshore oil by charging a 5 percent royalty, but this legal structuring had the unintended effect of creating a rush of lease applications by oil companies tired of the wildcatting competition. As charges of corruption and evidence of pollu-tion mounted, the state legislature was forced to take stronger action, plac-ing a moratorium on all new offshore lease sales in January 1929. By then the Standard Oil Company had developed slant-drilling technology, which allowed it to tap into state-controlled "submerged lands" from its onshore rigs in Huntington Beach (a practice later ruled illegal).

Louisiana was going through a similar oil boom in its southern swamps, lakes, and marshes, but with its thin coastal population and lack of recreational opportunities along much of the flood-prone Mississippi Delta, there was little resistance to the blowouts, fires, and other pollution taking place there. In fact, many of the area's settlers had always made their living through economic exploitation of the swamp, from fishing and trap-ping to market hunting of ducks and old-growth cypress logging, which resulted in the commercial extinction of the trees, just around the time the oil companies arrived.

"Oil provided an alternative," writes University of Southwestern Louisiana Professor Robert Gramling, "and the shift from one exploitative use of the region to another seemed natural and unproblematic." In the 1930s Gulf Oil and other companies, having developed drill barges for use in the swamps, began dredging hundreds of miles of canals to access their claims. These canals and associated erosion and subsidence became major contributors to Louisiana's subsequent land loss.

While it was becoming clear that onshore salt domes and other geological features associated with oil also extended offshore, the world refused to recognize national claims to economic resources beyond the 3-mile limit. But World War II changed all that.

As early as 1943 President Roosevelt, agreeing with Secretary of the Interior Harold Ickes, wrote that the 3-mile limit "should be superseded by a rule of common sense. For instance, the Gulf of Mexico is bounded on the south by Mexico and on the north by the United States. In parts of the gulf, shallow water extends very many miles offshore. It seems to me that the Mexican government should be entitled to drill for oil in the southern half of the gulf and we in the northern half of the gulf. That would be far more sensible than allowing some European nation, for example, to come in there and drill."

By the end of the war, seeing no naval powers likely to challenge U.S. claims, Ickes wrote a proposal that asserted the U.S. right to drill for oil on the submerged lands of the continental shelf, as well as to enter into fish conservation treaties with other nations fishing on the shelf.

The State Department strongly opposed this breaking of a legal precedent that went back to the Dutch lawyer Hugo Grotius's 1609 essay *Mare liberum*, which (with the backing of Queen Elizabeth and the British Navy) established the principle of open seas.

Still, Roosevelt's Secretary of State Edward Stettinius had bigger fish to fry with the upcoming Yalta Conference, which would determine how postwar Europe was to be broken up. As a result, Ickes was able to get a less experienced State Department official to sign off on his plan. In March 1945 a gravely ill Roosevelt gave final approval to the Ickes plan. Following Roosevelt's death, the atomic bombings of Hiroshima and Nagasaki, and the unconditional surrender of Japan that marked the end of World War II, President Harry Truman announced America's claim to its Outer Continental Shelf oil on September 28, 1945. This became known as the Truman Proclamation.

The political power of big oil was already well known, from the breakup of the Standard Oil trust and the Teapot Dome scandal to what became the first scandal of the Truman administration, resulting in the resignation of Ickes in February 1946. Ickes resigned to protest Truman's nomination of former Democratic Party treasurer Edwin Pauley as undersecretary of the navy. In that position Pauley would control the U.S. naval oil reserves. But Pauley was also known as a bag man for the oil companies and during the war allegedly told Ickes and Roosevelt that if the federal government did not challenge state claims to offshore oil (where the companies felt they would get a better deal), the oil companies would make major contributions to the Democratic Party. Ickes and other witnesses recounted Pauley continuing to push this idea on the train returning from Roosevelt's funeral. With the press hot on his trail, Pauley withdrew his name from nomination. Just over half a century later, the Clinton administration, having denied the oil companies drilling rights in the Arctic National Wildlife Refuge of Alaska, offered them access to the nearby National (formerly Naval) Petroleum Reserve.

By the end of World War II, the Texas and Oklahoma wildcatting oil-boom days were long gone. The postwar industry was consolidating its political and economic power, as graphically illustrated by the oil derricks pumping on the front lawn of the state capitol in Oklahoma City. This left a small independent company founded by former Oklahoma Governor Robert S. Kerr and his friend Deane McGee with little chance of securing any top-grade land leases. So, Kerr and McGee decided to gamble with some new technologies.

The end of the war had brought experienced navy men and surplus landing ships, or LSTs, home to America's coastlines. In 1946 the Magnolia Petroleum Company (later part of Mobil and then Exxon-Mobil), used navy veterans to begin a drilling operation in the gulf 5 miles off Louisiana on a platform made of wood and steel. Surplus navy ships housed the crew on the leeward side of a nearby island and shrimp boats transported them back and forth. Because it drilled a dry hole, history has tended to ignore Magnolia's breakthrough effort. Credit for starting the offshore industry usually goes to Kerr-McGee, which established a platform anchored by navy surplus ships 12 miles south of Terrebonne Parish in the fall of 1947. Two and a half weeks after it began drilling in 16 to 18 feet of water, a roustabout called his boss on shore on the morning of October 4 and told him oil was collecting in the drilling mud. "Well, skim it off," his boss replied. "Skim it off. Hell. There's barrels of it," the oil worker declared, heralding in one of the great new frontier energy booms in American history.

Conflicts between coastal states and the federal government over who could claim royalties from this bonanza escalated in the courts and in the press until 1953, when Congress passed the Submerged Lands Act, giving states control out to the 3-mile limit and providing for federal jurisdiction over all Outer Continental Shelf (OCS) submerged lands beyond.

With the rules firmly established, investors and speculators swarmed offshore, including the son of an eastern Senator, veteran war pilot and future president George Bush. Bush was among what *Fortune* magazine called the "swarm of young Ivy Leaguers" who descended on the "isolated west Texas oil town" of Midland in the early postwar years, anxious to make new fortunes apart from those they were already in line to inherit. Bush and his partners formed Zapata Oil, after the Marlon Brando movie *Viva Zapata!* that was playing in a Midland movie theater at the time.

George Bush later recalled, "Hugh Liedtke and I got in the offshore drilling contracting business and worked out a deal with LeTourneau and built three-legged rigs back in the mid-fifties. And they went on to become prototypes for drilling equipment that sat on the bottom. Our timing was good. We suffered a couple of setbacks. The first rig bent a leg or did something, had to be hauled into Galveston to be reworked. And the third or fourth one, the third one, disappeared in a hurricane, just vanished. I went out. I've never felt my eyeballs actually ache. I was flying in a single-engine plane with Hoyt Taylor. We were looking for any sign of it. We'd taken the people off and it was gone. A $6 million investment, that would be more like $76 million today. I loved the business. It was pioneering, the LeTourneau design, and also our people were pioneers. I felt we were in on the early stages of a marvelous business."

Offshore drilling would also prove to be a marvelous cash cow, generating more than a trillion dollars for the oil industry and over $125 billion for the federal government during the second half of the twentieth century. Today, offshore oil and gas royalties and lease sales provide around $6 billion a year in state and federal revenues. About a third of that goes to Louisiana, Texas, Alabama, California, and Alaska. The remaining $4 billion makes up the U.S. Treasury's second largest source of revenue after taxes (in close competition with customs tariffs).

There was some local resistance to drilling in the gulf, but it quickly gave way. *Thunder Bay,* a 1953 movie starring Jimmy Stewart as an oil exploration geologist confronting suspicious shrimp fishermen in Louisiana's bayou country, reflects the dominant view of the time when progress and industry were thought to be synonymous. It ends with the happy site of an oil gusher promising new wealth for the misbegotten Cajuns and a bright future for all. Today a gusher is seen as an ecological disaster, and Cajun and other regional identities are seen as valued heritages in our increasingly homogenized corporate culture.

By the late 1950s a number of shrimpers, including Edison Chouest, had converted their vessels to oil industry crew boats and began building the largest offshore fleet in the world. Edison Chouest Offshore is today best known for its specialty boats such as the Antarctic research vessel *Laurence M. Gould.*

The construction of an oil and gas infrastructure in the gulf also had unprecedented physical and environmental impacts with the appearance of fabrication yards, ship channels, roads, pipelines, and onshore production facilities, including 18 major oil refineries in Louisiana alone. Over the last generation, 136 polluting petrochemical complexes and six refineries located between New Orleans and Baton Rouge gave this 90-mile stretch of the Mississippi a grim but not inaccurate nickname, "Cancer Alley."

Another downside to America's marvelous new business became apparent on January 27, 1969, when a Union Oil platform off Santa Barbara, California, had a blowout. Less than a year earlier the Department of the Interior had sold drilling rights to 71 OCS tracts in the Santa Barbara Channel for $603 million. This was done over the objections of Santa Barbara residents, whose opposition to offshore oil tracked back to the turn-of-the-century Summerland spills. By the time Union Oil started drilling from its platform A in January 1969, it had gotten federal waivers reducing its primary ocean floor drill pipe casing to 15 feet and secondary casing to 238 feet. This was instead of the 500 feet and 861 feet normally required for oil spill prevention. On their fourth well the drillers hit a snag and suddenly oil and gas exploded up the pipe string. In a choking spray the crew managed to cap the well, but then oil began leaking below the shortened casings and through surrounding geologic fault lines. Within days some 3 million gallons had come ashore, turning 150 miles of beach to goo and ocean waves into a sludgy pulse.

"Get oil out!" became a local, then a state and national, rallying cry. Pictures of dying oil-soaked birds and the ineffective spreading of hay and cat litter onto the sand to soak up the oil inspired the beginnings of what would become a national environmental movement targeting industrial air and water pollution. President Richard Nixon came to look at the beach under the silent gaze of thousands of angry residents. For months afterward few people ventured onto Santa Barbara's fouled strand, except for a couple of hard-core surfers who were willing to follow up a session in the waves with a stinging turpentine bath, the only way to remove the sticky oil and tar balls from their skin, eyelids, and hair.

Then, on February 11, 1970, a Chevron rig in the Gulf of Mexico caught fire and began spilling oil. It burned for weeks before being blown out with explosives by famed oil firefighter Paul "Red" Adair. The platform was found not to have been equipped with a storm choke, a device designed to cut off the oil flow in an emergency. A subsequent investigation ordered by Secretary of the Interior Walter Hickel found that hundreds of rigs were operating without chokes. The feds indicted Chevron on 900 counts. The company was found guilty and fined $1 million.

In 1979 the Mexican-owned and U.S.-operated Ixtoc platform in the Gulf of Mexico exploded, gushing 150 million gallons of oil in a fiery uncontrolled spill (several men died trying to control it) that lasted 10 months and fouled the beaches of Texas, including the Los Padres

National Seashore. While some coastal communities were up in arms, the oil-dependent state government kept notably silent during the ongoing ecodisaster.

In 1982 Secretary of the Interior James Watt, less concerned about oil spills than charges of fiscal mismanagement, established the Mineral Management Service to administer offshore oil leasing and collect royalties. Through most of the 1980s Watt and his successor, Don Hodel (now cochairman of the Christian Coalition), attempted to lease a billion acres of the OCS for new oil development, arguing that if Congress prevented them from moving forward with their plan they would "be putting a sign on America that says we're willing to blindfold ourselves to our God-given resources and place ourselves at the tender mercies of OPEC."

Their attempts to lease huge expanses of the Blue Frontier, from fishery-rich waters off Alaska, New England, and North Carolina to the coral shallows of the Florida Keys and the chill waters off the redwood and pine bluffs of Mendocino, California, one of the most scenic stretches of coastline in the world, sparked massive protests and annual congressional drilling moratoriums that ended up excluding 85 percent of the U.S. coast from new leasing. The oil industry objected to these moratoriums, arguing that without more offshore platforms and pipelines the United States would become increasingly dependent on oil tanker traffic, which posed even greater risks of disaster.

In his 1976 ode to Alaska, *Coming into the Country*, author John McPhee recalls meeting local people worried about the Trans-Alaska pipeline then being constructed from the north slope's Prudhoe Bay oil fields to the port of Valdez. "There is the real problem," he quotes one local, "not the possible spills on land but the spills that could happen in Prince William Sound."

It was a fear repeatedly expressed (and ignored) by fishermen out of Cordova and other fishing ports on the sound. As early as 1971 they filed a lawsuit seeking to block the pipeline and what they called the "inevitable major" oil spill that would result from a tanker accident in Prince William Sound. They lobbied for laws that would require the oil companies to establish baseline environmental studies of the sound and implement spill-prevention and response measures, including tanker escorts, double-hulled tankers, and cleanup contingency planning, but by the time the tankers started sailing in 1977, few of these precautions were in place.

After years of oil industry cutbacks in tanker crew size and training, construction of fleets of very large and ultra-large crude carriers (VLCCs and ULCCs), and resistance to double hulling, the *Exxon Valdez* oil spill in Prince William Sound in 1989, caused when Captain Joseph Hazelwood, after drinking ashore, let an inexperienced crewman run his ship onto Bligh Reef, seemed a self-fulfilling confirmation of the warnings.

It is still hard to grasp the extent of the 11-million-gallon spill that resulted. Americans saw images of oil-soaked bald eagles, 1000 dead otters, a stunned and oil-blackened grizzly bear padding along a rocky shore, of moon-suited cleanup crews working futilely to scrub or steam soak 1200 miles of oil-covered wilderness coastline or skim 3000 square miles of water stained with toxic rainbow sheens. With the herring and salmon seasons ruined, fishermen and Native American communities faced futures of economic uncertainty and social dislocation. John Mitchell, an editor at *National Geographic,* described returning to these same communities ten years later and finding people still unable to get beyond the trauma of it, "just not able to let it go."

If a similar spill had occurred on the West Coast of the lower 48 states it would have stretched from Mendocino, California, 100 miles north of San Francisco Bay, to the Mexican border. On the East Coast, it would have spread from Cape Cod, Massachusetts, to below Cape Hatteras, North Carolina.

Still, every year less dramatic spills and discharges from offshore operations and coastal refineries, oily water illegally pumped from ship bilges, unburned gasoline from two-stroke outboards and personal watercraft, motor oil dumped down storm drains, leaking fuel from underground tanks, roadbeds, and parking lots along with other petrochemical runoff puts the equivalent of 20 *Exxon Valdez* spills into America's coastal waters.

Just before I visited the BP platforms there was a "minor" spill of 47,000 gallons of oil from a Chevron pipeline near Grand Isle, Louisiana, that created a 4-mile oil slick and fouled a small barrier island. On my first day offshore there was a natural gas blowout at the Apache Well platform, resulting in two injuries and lots of damage to the drill deck. A month later there was a 94,000-gallon oil spill that created a 7-mile slick 75 miles offshore when a drilling rig anchor was dragged across an underwater pipeline.

In the wake of the *Valdez* spill, Exxon agreed to pay a billion dollars into a state and federally administered oil spill trust fund. Much of that money has gone to purchase or protect half a million acres of wildlife habitat around Prince William Sound. In 1994 a federal jury in Anchorage awarded fishermen and affected communities an additional $5 billion in punitive damages, but six years later Exxon was still appealing that ruling.

The year after the Prince William spill, there were other major oil spills in Arthur Kill, New York, and off Texas, New England, and Huntington Beach, California. However, passage of the *Valdez*-inspired Oil Pollution Act of 1990 (OPA 90) had some salutary effects. Making shipowners responsible for damages and cleanup costs for oil spills and opening them up to unlimited liability in cases of gross negligence and violation of safety rules seemed to get their attention. In 1991 tanker spills in U.S. waters dropped to their lowest levels in 14 years. An industry-sponsored study documented improved safety inspections, awareness, and prevention procedures in the face of this tough new liability law.

Still, the main provision of OPA 90 called for double hulling of all oil tankers operating in U.S. waters by 2015. Recently, the oil industry began warning that 25 years may not be enough time to get ready (even though the average lifespan of an oil tanker is only 20 years).

"People are not knocking down our doors placing orders (for double-hulled tankers)," says American Shipbuilding Association President Cynthia Brown. "There's no sense of urgency, and if they don't order soon, shipyards won't be able to deliver on time. I think this is an intentional strategy by the oil companies to delay. And then they'll all order at once. And when the yards can't deliver, they'll ask Congress for relief."

The oil industry was certainly relieved in early 2000 when the Supreme Court ruled in favor of the International Association of Independent Tanker Owners (INTERTANKO) in their suit against Washington State. INTER-TANKO argued that states do not have the right to create tougher safety standards than those established under OPA 90. Under Washington State's "best achievable protection" regulations, vessel operators had to get approval for oil spill prevention plans from state authorities as well as the Coast Guard. The court found that federal laws take precedence, even where those laws may not be sufficient to safeguard the environment.

A second suit was brought in the U.S. Court of Claims by Maritrans, Inc., a tank barge company, arguing that OPA 90, by shortening the economic life of its single-hulled oil barges, was an unconstitutional taking under the Fifth Amendment. The company demanded a billion dollars in compensation.

The U.S. Court of Claims is a popular venue for "regulatory takings" claims by corporations that do not want to comply with maritime safety regulations, coastal developers who are prevented from building in hurricane zones, and mining companies that are denied permits to dig up salt marshes.

Part of the Fifth Amendment states, "No person shall be . . . deprived of life, liberty, or property, without due process of law; nor shall private property be taken for public use without just compensation." Today, when the government condemns land to build a highway or commercial port, the Fifth Amendment guarantees that the landowner be paid market value for lost property.

In 1887 the Kansas beer brewer Peter Mugler argued the first case for a regulatory taking, claiming that a prohibition law passed in his state was a taking under the Fifth Amendment because it devalued his property, putting him out of business. The Supreme Court ruled against him, stating that "a government can prevent a property owner from using his property to injure others without having to compensate the owner for the value of the forbidden use." This "nuisance clause" is the basis on which the government has been able to establish health and safety regulations, zoning, labor codes, and consumer and environmental standards.

What more than a century of this precedent has tended to show is that for every form of regulatory taking, there are far more "givings" going on. The land-use plan or taking that prevents a corporate hog farm from dumping manure into an upstream river may be seen as a "giving" by the coastal resident whose beachfront is not suddenly smothered by a bacteria-rich, nutrient fed algae bloom and fish kill. But today's takings advocates in the oil industry are not looking for legal precedent to advance their positions but to the political and judicial legacy of the Reagan administration.

In 1985 Richard A. Epstein, a professor at the University of Chicago Law School, wrote *Takings: Private Property and the Power of Eminent Domain,* in which he argued that the Fifth Amendment requires the government to pay property owners compensation whenever regulations or laws limited the value of their property. He goes on to claim that along with environmental laws and building permits, income taxes are also a form of takings. That Epstein's eccentric interpretation of the Constitution found favor in Washington was confirmed by Charles Fried, the U.S. solicitor general under President Ronald Reagan from 1985 to 1989, who wrote, "Attorney General Meese and his young advisors . . . often devotees of the extreme libertarian views of Chicago Law Professor Richard Epstein—had a specific, aggressive, and, it seemed to me, quite radical project in mind: to use the takings clause of the Fifth Amendment as a severe brake upon federal and state regulation of business and property."

Reagan appointee Loren Smith, the chief judge of the U.S. Court of Claims, still numbers Epstein among his philosophical heroes. Several members of the court share Smith's views, which is what attracts takings plaintiffs to their chamber like lobsters to a herring-baited lobster pot.

Of course, it still remains far easier for the oil industry to score extra billions from Congress than from the courts. Despite growing concern over climate change and its links to the burning of fossil fuels, Congress continues to provide some $12 billion a year in federal subsidies to the oil industry, not counting the half-billion-a-year royalty holiday for deep-water drilling that ran between 1995 and 2000.

To keep this federal largesse coming, the oil and gas industry is willing to prime the pump, spending more than $80 million a year inside the Washington Beltway. In 1998 it spent $57,696,393 on K-Street Washington lobbyists and passed out $22,007,424 in direct federal campaign contributions. That same year it helped kill the American Oceans Act, which would have set up a national commission to study the state of America's Blue Frontier.

"The petroleum industry presence was definitely felt on the Ocean Act," says Jim Saxton, the Republican representative from New Jersey who was one of the bill's cosponsors. "We got the bill out but it never got through the Senate. Essentially, its aim was to try and find a better way to do ocean policy."

"So why would the oil industry oppose that?" I wonder.

"An industry doesn't like rules changed when they feel like they're getting what they want. They had their soldiers here in Congress doing their work. I'd ask them what they objected to in the bill, and they'd keep saying one thing and then another and soon it became clear they were just out to scuttle the bill. They just like the way they're doing business now." (The Oceans Act was finally passed in July 2000.)

I get to see an example of how "they're doing business" on September 29, 1995, inside U.S. Capitol Room H-137, located off a narrow tourist-clogged hallway, not far from the Rotunda. On the meeting room's blue and white carpet, folding chairs and tables have been formed into a square for a House-Senate conference. An outer ring of chairs is rapidly filling with staffers, oil lobbyists, and reporters. Alaskan Republican Representative Don Young, chair of the House Resources Committee, presides over the meeting. He's wearing a dark suit, paisley tie, and what looks like a red .22-caliber round as a lapel pin. "It's actually a miniature shotgun shell," the lifetime NRA member later tells me, "to signify I'm a bird shooter. I shoot doves and quails and all those good things."

The conference is about granting the oil companies that half-billion-dollar-a-year royalty holiday, to encourage the technological innovation needed for deep-sea drilling. It's an idea being promoted by soon-to-retire Democratic Senator Bennett Johnston of Louisiana. After retirement, Johnston will set up his own multimillion-dollar Washington lobbying firm representing various energy companies and alliances. The proposal also has the support of the Senate Energy and Natural Resources Committee chairman, Republican Frank Murkowski of Alaska. It's being pushed as a rider to a bill allowing the export of Alaskan crude to Asia.

George Miller, the 6-foot 4-inch, white-haired Democrat and fierce environmental advocate from California, claims that granting the royalty holiday "is like throwing rose petals after the parade has gone by." Industry has already developed the deep-water technology it needs for drilling, he argues. (Shell drilled Cognac, the first platform in more than 1000 feet of water, way back in 1978.) "The oil executives can make some serious money and don't need this larding on," Miller complains. Turning to Young, he asks, "Why are we giving taxpayer money to these people who have no need for it?"

Young turns to Johnston and says, "Very frankly, senator, I'm not happy to have this rider on the bill." Still, Young goes on to vote for it along with a majority of the House-Senate conferees.

"Let's remember we fought a war a few years ago over one thing and that was oil," Murkowski explains to a small gaggle of reporters after the meeting adjourns. "We fought the Gulf War for oil?" I ask. "I didn't say that," he counters.

The five-year deep-water royalty holiday expired on November 28, 2000. The National Ocean Industries Association (NOIA), which represents the offshore industry, decided not to lobby for a renewal in the midst of their hugely profitable deep-water drilling boom. "I'm not interested in chasing a lost cause up on the hill," says NOIA president Robert Stewart before going on to justify the $2.5 billion government subsidy the industry had received, claiming that it "allowed, maybe that's too strong . . . let's say contributed to, an enormous increase in deep-water leasing."

"The number of leases went up with the Royalty Relief Act and it contributed to that," agrees Hugh Depland, who, aside from his job with BP, is also public relations chair for NOIA. "The technology was moving forward independently, but this brought some new players into deep water, and certainly was of some assistance to us." I understand. It certainly would be of some assistance to me if I could get Congress to pass a rider to a bill giving me a major tax cut to encourage literary innovation and reduce America's dependence on foreign authors.

"We project 50 percent of total offshore production will be from deep waters in the next year or two," says Jim Regg the thin-faced, bearded chief of technical assessments and operations for the Mineral Management Service in the Gulf of Mexico. The MMS gulf offices occupy nine floors of a 10-story high-rise next to a K-Mart in a suburban New Orleans shopping mall.

"That's 50 percent of all the oil and gas?" I ask.

"They're getting more oil than gas right now but we don't think that's a geological factor in deep water, it's just a rampant rollout of oil production as they chase the larger income sources, just the normal trend to go after the oil first. BP-Amoco, Shell, Exxon are all selling off their shallow-water holdings [and going deep]. We have the potential of seeing 10,000-foot wells drilled soon. The *Glomar Explorer* has that potential. So does the *Pathfinder* and *Millennium*. Those three drill ships are moving in and out of the gulf, moving between the gulf, West Africa, and Brazil."

I ask him about the effects of future oil spills that will inevitably occur in deep water. NOIA's Robert Stewart had admitted that no one knows much about what happens when oil is released in that kind of extreme high-pressure environment, how it might spread, or what could be done to recover it.

"We're working on a deep spill project," Regg tells me. "It's been going on for two years now. There's a lab in Hawaii with high-pressure chambers to recreate the water pressures at depth. We've also got a planned release in the North Sea to see if the oil disperses and how, and also if the natural gas goes into a hydrate [jelly-like] phase or what. There are huge unknowns. We're cofunding the study with industry. The International Association for Drilling Contractors has established some guidelines for deep-water spills."

I ask about the Coast Guard's role as the lead agency for oil-spill response. "The Coast Guard's involved in all the reviews," Regg assures me. "We have lots of meetings with industry that they sit in on."

Since James Watt created MMS in 1982, the agency has never canceled a lease sale based on an oil-spill risk assessment. It is also clear that it's not about to slow down today's deep-water drilling boom just because no one has any idea how to respond when the first deep-water spill occurs. For formality's sake I ask Regg if MMS has ever canceled a lease to protect any part of the marine environment.

"Some leases have been canceled because of politics," Regg points out. He refers to a planned lease in the eastern gulf off the Florida Panhandle (lease sale 181 scheduled for December 2001) as "our California." What he means by that is there's widespread public opposition to the sale, even though MMS and the oil industry are anxious to see it move forward.

"Is there any program in MMS to encourage the production of natural gas over oil?" I wonder.

"No, nothing like that," he says, seeming perplexed.

Which is too bad, given that with less than half the carbon dioxide content of petroleum, many climate experts have begun viewing natural gas as the logical transition fuel between oil and carbon-free renewable energies like solar photovoltaics, wind turbines, and hydrogen fuel cells (which strip combustible hydrogen molecules from alcohol and other fuels, leaving only water behind).

Natural gas is an in-place part of the world's energy infrastructure and so would require relatively little capital outlay to bring it online as a petroleum substitute. It is already the power-plant fuel of choice in Europe. Concentrated as a liquid, natural gas is also volatile enough that it quickly evaporates when spilled. This would provide the added benefit of eliminating oil-spill impacts on America and the world's marine ecosystems, something that can never be assured as long as massive amounts of oil and saltwater remain separated by mere inches of metal piping and tanker hull.

Interestingly, John Browne, BP's CEO, was the first leader of a major oil company to acknowledge that the science on climate change is sound. Looking to the future, BP has even begun calling itself an energy company rather than an oil company.

"We see a shift to lower carbon energy," Hugh Depland tells me. "The world is moving from coal to oil to natural gas and then to carbon-free hydrogen."

Right now BP is America's largest producer of natural gas, but it is also America's largest producer of oil. BP has invested a billion dollars in solar energy but has also spent $26 billion to acquire ARCO, making it the second largest "energy company," after Exxon-Mobil.

In addition, BP has pledged to reduce carbon dioxide emissions from its oil refineries by 10 percent, which is kind of like pledging to reduce gun violence at the Colt firearms factory. Whatever reductions in emissions it does make in refining will be more than offset by its massive Northstar drilling project, the first offshore oil operation (with under-ice pipelines) planned for Alaska's Arctic Ocean.

I ask Hugh Depland if BP has a preference for drilling natural gas over oil. "We have a preference for whatever's economically advantageous," he admits.

The truth is that market forces are wonderful tools for doing things like making sure we don't run out of gasoline during the holiday-driving season. But for responding to global threats like Adolf Hitler, nuclear war, or global warming, we still need government. What BP has done is to recognize that with the world entering its most challenging environmental crisis in history, those who come to the negotiating table early will have a greater say. Still, it remains up to the world's political leadership to set the course for a transition from fossil fuels to new, cleaner forms of noncarbon energy.

For now, climate change is an issue that does not seem to affect most Americans in their daily lives the way health care, education, or even the Internet might. When you do think about climate, it is usually during one of those increasingly common record-breaking heat waves, and then you might respond by turning on the air conditioner in your car or sports utility vehicle.

Still, for the majority of Americans who live in coastal regions, there are more extreme climate impacts coming to the Blue Frontier that, once understood, might really make you sweat.

[Chapter Six] A Rising Tide

[On that day] were all the fountains of the great deep broken up, and the windows of heaven were opened.

—GENESIS 7:11

I spoke to a man on Wall Street who was stuck in his office looking down as his Porsche sank. I asked him what he did. "Wait for the tide to go out," he said.

—VIVIEN GORNITZ, NASA SCIENTIST,
RECALLING THE NOR'EASTER OF 1992

Clouds, snow, rock, and water hard as black marble is all we can see approaching the end of our four-day, 900-mile journey from Punta Arenas, Chile. Up on the bridge of the *Laurence M. Gould,* Robert, the first mate, is playing Led Zeppelin and talking on the radio with Palmer Station. "Never been this far south without seein' ice," he says in a lilting Cajun accent.

"That's 'cause we cleared it for you," the base's radioman jokes. "Went out in our Zodiacs with blowtorches." We round Bonaparte Point, and there it is, set in a boulder field below a blue-white glacier—Palmer Station, Antarctica.

Palmer, one of three U.S. Antarctic bases run by the National Science Foundation, is where I got to spend six weeks one recent austral summer (the northern winter of 1999). Palmer is located on Anvers Island, 38 miles of granite rock covered by ice up to 2000 feet thick. Anvers is part of the Antarctic Peninsula, a 700-mile-long tail to the coldest, driest, highest continent on Earth, a landmass bigger than the United States and Mexico combined, containing 70 percent of the world's freshwater and 90 percent of its ice. The peninsula, where polar and marine climates converge, is also a wildlife-rich habitat that researchers refer to as "the banana belt." And that was before global warming.

While docking, we're greeted by a small welcoming committee of people, Adélie penguins, skuas, and elephant seals. Palmer Station has the look of a low-rent ski resort next to an outdoor equipment dealership. It is made up of a group of blue and white prefab metal buildings, with two big fuel tanks, front loaders, snowmobiles, and Milvans scattered around. The two main buildings, Biolab and GWR (garage, warehouse, and recreation), are separated so that if one burns down the other can act as a refuge for the 20 to 40 scientists and support personnel who work here year round.

The short oblong pier on the inlet has giant rubber fenders, where the *Gould,* our 240-foot supply and research vessel, docks every six to eight weeks during the summer. January and February's summer temperatures drift between a balmy zero and 40 degrees Fahrenheit, with 23 hours of daylight to enjoy the views. The weather is variable, with sun, clouds, wind, rain, snow, and gale force winds, often on the same day, kind of like the San Francisco Bay area on steroids. Next to the pier is the boathouse and its string of black and gray Zodiacs, Mark 3s, and Mark 5s. Fifteen to 20 feet in length, we'll get to operate these fast rubber rafts in the surrounding island-studded subfreezing waters on days when the winds drop below 20 knots.

Since 1970 climatologists have predicted that global warming, as it kicked in, would occur most rapidly at the poles, a fact now confirmed by scientists in Alaska, Canada, Greenland, at the North Pole, and here on the Antarctic Peninsula.

Bill Fraser, a rangy, sun-weathered, 48-year-old ice veteran from Montana State University is the chief scientist at Palmer. "The Marr glacier used to come within 100 yards of the station," he tells me, pointing upslope. "Its melt water was the source of our freshwater." Today the Marr is a quarter-mile hike from Palmer across granite rocks and boulders. Skua birds now splash in the old melt pond while the station is forced to use a saltwater intake pipe and reverse osmosis desalinization to generate its freshwater. Periodically, the artillery rumble and boom of moving ice alerts us to continued glacial retreat and allows for spectacular views of irregular ice faces collapsing into Arthur Harbor, setting off a blue pall of ice crystals and a rolling turquoise wave beneath a newborn scree of chunky brash ice.

"When I was a graduate student we were told climate change occurs, but you'll never see the effects in your lifetime," Bill says. "But in the last 20 years I've seen tremendous effects. I've seen islands pop out from under glaciers, I've seen species changing places and landscape ecology altered."

While global temperatures have warmed an average 1 degree Fahrenheit over the last century—paralleling increased industrial output of carbon dioxide and other greenhouse gases—the Antarctic Peninsula has seen a jump of more than 5 degrees in just 50 years, including an incredible 10-degree average warming during its winter months.

One way that we know there is more carbon dioxide in our atmosphere today than at any time in the past 420,000 years is through ice core samples taken from Siple Dome, Vostok, and other sites in the Antarctic interior. These cores contain trapped bubbles of ancient air that have been isolated, dated, and chemically analyzed. They also show that climate is far less stable than we've imagined, and that the past 10,000 years—the period that has seen the rise of human civilization—has also been a period of atypical climate stability.

The Antarctic Peninsula made news recently as huge pieces of the Larson-B ice shelf, including one iceberg twice the size of the state of Delaware, began calving off its eastern shore. Scientists are now discussing the possibility that the Western Antarctic ice sheet adjacent to the peninsula could experience a sudden meltdown, raising global sea levels by 18 to 20 feet (instead of the 1 to 3 feet currently predicted by 2100). While most experts believe this melting will occur sometime after this century, by the time they know for sure it will be too late to do anything about it.

So I can argue that I'm learning a survival skill when the first thing I do after helping unload the ship's "freshies" (fresh fruits and vegetables for the station) is learn how to operate one of the Zodiacs that, along with thick Sorel boots and ice crampons, provide the main means of transport at Palmer.

Taking the 15-foot rubber boat through floating fragments of brash ice, I spot a leopard seal lazing on an ice floe. I maneuver around to take some photos of the snaky, blunt-headed predator, when a panicked penguin jumps into the boat, tripping over the outboard's gas can. We exchange looks of mutual bewilderment before it leaps onto a pontoon and dives back into the icy blue water.

A few days later I'm out with Bill Fraser and his "Schnappers" (the boat-radio moniker for his seabird researchers, in honor of a Wisconsin polka band). We tie off our bowline on the rocky edge of Humble Island. Removing our orange float coats, we walk up to a wide pebbly flat past a dozen burbling 1000-pound elephant seals lying in their own green waste. One of them rises up just enough to show us a wide pink mouth and issue a belching challenge that means stay back or I might have to rouse myself from complete stupor in order to attack you. The elephant seal population, once restricted to more northerly climes, is now booming along the peninsula because of warming conditions.

Their belching, grunting sounds are soon complemented by the hectic squawking, flipper-flapping, and cow-barn odor of 3000 Adélie penguins and their downy chicks, who occupy a series of rocky benches stained the color of Georgia red clay by their krill-rich droppings. Brown gull-like skuas, looking for a weak chick to kill and feed on, glide majestically overhead.

"These penguins are the ultimate canaries in the mine shaft. They're extremely sensitive indicators of climate change," Bill tells me as we walk past a group of 2-foot-tall adults waddling up from the sea, their bellies full of krill.

Tiny shrimplike creatures, krill are the most abundant animal on earth in terms of their total biomass. They make up the broad base of Antarctica's food chain and are consumed in vast quantities by penguins, seals, and whales (a single blue whale can eat four tons a day). But without access to sea ice, krill shrink, lose weight, and are vulnerable to early death.

"The bottom of the ice is where 70 percent of krill larvae are found," explains Dr. Robin Ross of the University of California at Santa Barbara, a prim and cautious scientist who'll spend the research season onboard the *Gould* trolling for krill and plankton. "The ice is like an upside down coral reef with lots of bumps and crevasses and caves for them to hide in," she continues. "But in the early nineties the cycles of high and low ice began to fall apart. This year's winter sea ice was the lowest on record."

This year's trawls are also bringing up more salps than krill. Salps are open-water jelly creatures called tunicates, which look like floating condoms, foul the bow lines of our Zodiacs, and are prey for only a limited number of birds and fish. Unlike krill, salps reproduce in open water—and may soon fill the ecological niche created if declining sea ice leads to a long-term decline in krill. A decline in krill would wreck much of Antarctica's living ecosystem.

Rising temperatures also increase precipitation, which in Antarctica takes the form of snow. Excessive spring snow has disrupted the nesting and breeding of Adélie penguins, leading to the extinction of many of their island colonies.

Back on Humble, I'm waving off a dive-bombing skua while Bill is conferring with Rick Sanchez of the U.S. Geological Survey. Rick is carrying a portable GPS (global positioning system), along with a satellite antenna sticking out of his backpack and a magnesium-shelled laptop strapped to an elaborate fold-down rig hanging from his waist and shoulders. He's trying to walk off the perimeter (he calls it the polygon) of an extinct colony of Adélies in order to confirm Bill's observations linking increased snowfall to their declining numbers, but a burbling pile of elephant seals is blocking his mapping venture. If he tries to move them, they might stampede and crush still-living penguin chicks. Such are the quandaries of high-tech research projects in Antarctica.

While Adélies are going extinct, more adaptable species like chinstrap penguins (that eat fish and squid when krill are not available), elephant seals, and fur seals are increasing their numbers. These newcomers to the area are threatening to displace sea ice–dependent animals like Weddell seals, crabeater seals (which are actually krill eaters), and leopard seals (which also eat krill, penguins, crabeaters, and the occasional Zodiac bumper).

What these changes in the Antarctic Peninsula suggest is that rapid warming could speed up a global chain reaction of extinctions that—thanks to the impact of humans—is already underway and has been labeled the sixth great "extinction pulse" in planetary history (the last one, a meteor-based event, took out the dinosaurs some 60 million years ago).

In a warmer world, "weedlike" species that are highly adaptable to disrupted habitat (pigeons, rats, raccoons, deer, elephant seals, and chinstrap penguins) will displace more specialized endemic creatures (tigers, monarch butterflies, river dolphins, sea turtles, and Adélie penguins) that depend on unique ecosystems such as coral reefs, rain forests, and the Antarctic ice shelf. Rising temperatures may also kill off certain plant species.

Tad Day, a sandy-haired, boyish, 39-year-old professor from Arizona State, who drives his Mark 3 Zodiac "Luceille" like a Formula One racer, has been studying Antarctica's only two flowering plants, hair grass and pearlwort. The main study site he and his "Sundevils" use is Stepping Stone Island, a surprisingly green, rocky isle several miles south of Palmer around the rough chop of Bonaparte Point. "Step" is surrounded by pale blue icebergs, a rumbling blue-white glacier, and other rocky islands and outcroppings—including Biscoe Point to the south, which, with the retreat of the Marr glacier, has now become Biscoe Island. Amidst nesting giant petrels, albatrosslike scavengers the size of eagles, and a friendly skua named Yogi, Tad maintains two gardens, fenced to keep fur seals out, containing more than 90 wire plant frames surrounding banks of hair grass and pearlwort growing not in true soil but a close approximation made up of glacial sand and guano.

Day has found that warming improves the growth of pearlwort but appears to have a negative impact on hair grass, the dominant species in Antarctica that is now being displaced by pearlwort, a mosslike plant. "Global warming," he explains, "has the capacity to shift the competitive balance of species in ways that, until we get out there and do the research, we don't understand yet, and that could have important consequences on our ability to produce food and fiber."

Increasingly reliable climate models now predict a 3- to 11-degree planetary warming this century. This will result in shifts in agricultural production, spread of tropical insects and diseases, increases in extreme weather events, more intense coastal storms and hurricanes, erosion of beaches, coral bleaching, and rising sea levels—all of which have already begun.

Still, it's hard to maintain a sense of gloom and doom on the last wild continent, at least for more than a few hours at a time. Along with nightly discussions over Pisco sours with glacier ice at the Penguin Pub (the open bar located above the machine shop in GWR), I manage to distract myself with simple sojourns on the southern ocean. You need at least two people with radios to take out one of the Zodiacs, so on the days when I'm not working with the Schnappers or Sundevils I spend time looking for a boating partner. Doc Labarre, the station's big, balding, fatalistic physician who used to work in the emergency room in Kodiak, Alaska, is among those regularly up for an adventure.

One day we cruise past Torgersen Island, where I take the Zodiac up "on platform" (as you speed up the bow drops down, giving you greater visibility and control) and head us toward Loudwater Cove on the other side of Norsel Point. The following seas allow us to surf the 15-foot craft past the rocky spires of Litchfield Island and around the big breaking waves at Norsel. We then motor round a few sculptural apartment-sized icebergs, crossing over to a landing opposite the glacier wall. We tie off our bow line, watching a serpentine leopard seal sleeping on an adjacent ice floe. We dump our orange float coats and climb several hundred feet up and over some rocky scree and down a snow field splotched with red algae to the opening of an ice cave. In the cave it's like a dripping blue tunnel with slush over a clear ice floor that shows the rocky piedmont below. Hard blue glacier ice forms the bumpy roof with its stalactite-like icicles and delicate ice rills forming pressure joints along its edges.

Outside we hike the loose granite, feldspar, and glacial sand until we encounter a fur seal hauled up several hundred yards from the water on the sharp broken rocks. He barks and whines a warning at us. Nearby ponds and 100-year-old moss beds have attracted crowded colonies of brown skuas, who soon begin dive-bombing us. I get whacked from behind by one of the 5-pound scavengers. It feels like getting slapped hard in the back of the head by a large man. We quickly move away from their nests, climbing back over the exposed glacier rock, passed middens of limpet shells and down a rock chimney to where our boat is tied up. The leopard seal is awake now, checking us out as we take off. I notice blood stains on the ice where he's been resting.

We next drop by Christine, a big bouldery island where we walk past a large congregation of elephant seals hanging out opposite a colony of squawking Adélies. Crossing the heights we find mossy green swales with ponds full of brine shrimp. We then lay out on a rocky beach at the end of a narrow blue channel, sharing the space with two elephant seals about 500 and 900 pounds each. The southern ocean is crystal clear; the sun has come out and turned the sky cobalt blue. It feels almost tropical, lounging to the sound of the waves rolling and retreating across smooth fist-sized stones. Further out are several flat islands with big breakers crashing over them, sending spray 50 feet into the air. The elephant seals are blowing snot and blinking their huge red eyes, their black pupils the size of teaspoons for gathering light in deep-diving forays after squid. A fur seal comes corkscrewing through the channel's water before paddle walking ashore and scratching itself with a hind flipper, a blissful expression on his wolfy face. And there we are, just five lazy mammals enjoying a bit of sun.

Driving the Zodiac back to the station we're accompanied by a flight of blue-eyed shags and squads of leaping penguins in the water. Doc steers while I keep an eye out for whales, like the minke that bumped the boat I was riding in a few days earlier (a real Melville moment, watching its huge brown back roll out from under us). The sky has again quilted over with clouds, turning the water the color of hammered tin; with the buck and slap of the boat and the icy cold saltwater spray in our faces, it feels like all's right with the wild.

Antarctica is vast and awesome in its indifference to the human condition. At the same time it is a world center for scientific research and has provided us fair warning about the human impact on climate change. The message from the ice is as plain as the penguin bones I found scattered around a dying Adélie colony. Our world, and theirs, Antarctica's Southern Ocean and America's Blue Frontier, are more closely linked than we imagine.

Two weeks later I'm standing under tall and stately palm trees with members of the Alliance of Small Island States from Fiji, Samoa, and Micronesia. They're concerned that rising sea levels might submerge their tropical nations.

Soon another island resident arrives to lead us on a tour and explain how her island, which has seen its surrounding waters rise a foot in the last century, will be impacted by an additional 2- to 3-foot sea-level rise predicted for the new century.

Dr. Vivien Gornitz, a diminutive, plainspoken scientist and native New Yorker from NASA's Goddard Space Institute at Columbia University, leads the group out of the World Financial Center's Palm Court Atrium and over to lower Manhattan's Hudson Pier to take a ferry ride to Hoboken, New Jersey. The group is followed by a television camera crew and producer Andrea Torrice of Norman, Oklahoma, who is shooting a PBS documentary, *Rising Waters,* about climate change impacts on Pacific islanders.

Gornitz points to the green algae tidal marks halfway up the seawall that runs below Battery Park City. "This area was pretty much under water during the 8-foot storm tides of the big nor'easter of 1992," she explains. "I spoke to a man on Wall Street who was stuck in his office looking down as his Porsche sank. I asked him what he did. 'Wait for the tide to go out,' he said. In the coming century, 100-year storms like that one will begin to occur every decade or so. In areas like lower Manhattan they'll raise the sea-walls. There's too much valuable real estate here to let the sea take it. Of course that's not feasible for parts of Long Island, North Carolina, Florida, or Louisiana."

Aboard the ferry Torrice and her crew herd the tour group onto the exposed upper deck, into the teeth of a cold March wind. "But we'll freeze up there," Gornitz protests.

As the boat pulls away, affording a stunning view of the Statue of Liberty and the city, climatologist Penehuro Lefale of Apia, Samoa, takes a look back at the World Financial Center. "You have some really amazing architecture here," he tells me, smiling. "That building alone has to be worth more than the budgets of all our small island nations."

Dr. Gornitz, who lives in upper, dryer Manhattan is part of a National Science Foundation–funded team assessing climate change impacts in the metropolitan tristate area, part of a national assessment of climate change impacts on the United States coordinated by the White House Office of Science and Technology. She has focused on lower Manhattan, Sea Gate and Coney Island in Brooklyn, Westhampton and Freeport on Long Island, and parts of New Jersey, trying to go from an understanding of the macro to the micro effects of climate change. In Hoboken she takes the group through the marine terminal and down the stairwell to the commuter PATH trains.

"During the '92 storm these stairs became waterfalls as the river surged in," she explains, pulling out photos to illustrate.

She then shows a map of how a 10-foot surge, typical of the more extreme storm events predicted for this century, could flood all of New York's subway tunnels and three major airports.

"Here the waters cover the concrete. At home they flood our Taro fields and take away our food plants and houses," says UN ambassador from Micronesia Masao Nakayama, before recounting how during the big

cyclone of 1991 people on Truk had to get in their canoes and paddle as the
storm surge covered the entire island.

Asked about some white dome structures on the ride back to
Manhattan, Gornitz explains that Chelsea Piers has converted from a shipping dock to a health club. "Gyms and tennis courts under those domes. I
couldn't tell you exactly. I've never gone there to work out."

The old pier pilings are also working out as habitat for juvenile flounder and striped bass as the water quality of the river and harbor improve, a
phenomenon marked by the return of abundant fisheries and harbor seals.

Soon the group is back on the riverfront esplanade, walking south toward
the vulnerable areas of the South Street Seaport and lower Wall Street.

"Are people who live here aware of the sea level problem?" Penehuro
Lefale wonders.

"Not for the most part," Gornitz admits. "We live on an island in
Manhattan but we're cut off from the sea in our daily lives and don't think
about the water all around us. The irony is just as we're beginning to redis-
cover our waterfront and rebuild on it, it's at risk."

Penehuro, who lost his family home and grandmother's grave to a sav-
age storm, nods his understanding. "For us the sea and the land are very
much about who we are."

"That's great," Torrice interrupts. "But could you both say that again?
We had a problem with the camera battery."

A month later I'm back in New York for a follow-up meeting with
Gornitz. The NASA Goddard Institute for Space Studies at Columbia is
located in a university building above Tom's Restaurant, an easy-to-spot
landmark at the noisy, traffic-congested corner of Broadway and 112th
Street.

Gornitz's surprisingly quiet office is off one of the building's warren-like
hallways on the fifth floor. Wearing a mousy brown dress she appears
uncomfortable with this ongoing media interest. "I don't know what I can
tell you," she tells me. "We're working with the Army Corps [of Engineers]
in terms of their beach nourishment programs in New Jersey and how rising seas will impact that. When we have another storm like the '92 nor'easter,
those beaches will go out to sea again. Places like Sea Gate [Brooklyn] and
Sea Bright [New Jersey] are so vulnerable that reporters just head right
there in a storm knowing they'll have good flood footage. The corps people
think this is a problem to worry about 30 to 40 years down the road, but I
tell them it's already happening. The last two decades we're way above normal variability for climate change. We know 70 percent of sandy beaches
around the world are eroding. Global coral bleaching was at an all-time
high in 1998 [as sea temperatures warmed]. Oceans are rising 1 to 2 mil-

limeters a year according to the IPCC [the UN's Intergovernmental Panel on Climate Change]. The rate of sea-level rise over the last century is much faster than over the last several thousand years. Places already seeing sea-level rise impacts include Chesapeake Bay, Florida, and Louisiana. You can go there and see degradation of coastal wetlands and lost islands."

Does she see a lot of people taking note of these dramatic changes?

"If we keep getting storms like the '92 nor'easter, we'll start to take note. A lot of development in the last thirty years has taken place during a period of little storm activity."

Just off 117th Street in Miami, inside the entry gate to Florida International University, is a low reinforced-concrete and steel building with roll-down hurricane shutters and an array of satellite dishes and microwave masts rising like palmettos from its fortress flat roof. If the antennas come down in a storm, there are two buried fiber optic cables that will keep the operational core of the building connected to distant weather stations with uplinks to geosynchratic NOAA and navy satellites. All the building's entrances are under 24-hour video surveillance. The structure itself, completed in 1995, is elevated on a dirt berm, 5 feet above the surrounding campus. I figure this has to be the right address for NOAA's National Hurricane Center (NHC). It is also NOAA's Tropical Prediction Center and the Miami Weather Forecast Office, but their directors will not get to retire as network television consultants like Bob Sheets and Neil Frank, both former NHC directors, have done. I get buzzed in through a sealed metal door.

The NHC press room is the building's last protective ring before the core operations center. It has glass doors leading into the ops center, a briefing podium, and a producer's console for live video feeds. Just beyond the glass is a large TV monitor and desk facing outward toward the press room. "We use this as a TV set with the director sitting up front, explaining things to the pool camera," Frank Lapore, the center's press guy, explains to me. Behind the big props the ops center is divided between eastern Pacific and Atlantic sectors shown as computer-generated maps on various terminals and monitors. Hurricane specialist Dr. Jack Bevin is seated with a cup of coffee, keeping an eye on the Pacific this morning, as Dr. James Franklin works storms in the Atlantic.

Off to the right and out of view of the press room is the Prediction Center, where the Tropical Analysis and Forecasting Branch develops aviation and marine forecasts, churning out 50 charts a day for, among other clients, helicopter companies like Mark and Cathal's, which make daily flights to and from oil platforms in the Gulf of Mexico. "We fax these charts to the world. We're now online and getting 1.5 million hits a day, seeing exponential growth in product demand," Frank brags. I notice the

Weather Channel is playing on a TV monitor suspended from the ceiling. "We want to make sure our information is getting out and being presented accurately," he explains. "The media, particularly in this market, is very hyper. We try and act as the voice of reason."

A polar satellite display on one computer screen shows swaths of different colored earth, another shows swirls of red, gray, and blue, the blue representing the cold tops of clouds. Still another shows aviation zones in the Caribbean with country map outlines overlaid. I pick up a "25 July 1999 Ash Advisory" statement from Montserrat, a small island with ongoing volcanic eruptions. "That's important. Can't fly a jet well with its engines full of silicate," he notes.

We walk back into the ops center. "Must be a slow news day. They called asking if this low could be a storm in the making," Jack Bevin reports, pointing to one of his monitors as he hangs up the phone.

"Any stray cloud will do now that they're done burying [John] Kennedy [Jr.]," Jim Franklin smiles wanly. It is July 1999, and the news has been full of the Kennedy scion's plane crash and recovery, turning a human tragedy into another ghoulish ratings grabber.

"During a hurricane we might have half a dozen people working in here and a couple of dozen straphangers including the tribal elders hanging around, imparting their wisdom. Everyone is pumped up. It's exciting. You should come back during an event," Frank offers.

I consider this when the 600-mile-wide Hurricane Floyd, with internal winds of over 145 miles per hour, looks like it's about to smooth out Florida's rough edges and close to three million people are hitting the roads of the southeastern United States in the largest civilian evacuation in U.S. history. By the time I call the airlines to inquire about a flight, however, access has been shut down. Then the September 1999 superstorm veers north.

"We packed up our dogs and guns and silver and put the boat on the trailer and took off," Deb Sheppard, a coastal resident of Darien, Georgia, recalls. "We have a marsh-front house on the coast. We took pictures of all our possessions and didn't look back. Figured if we lost the house we weren't going to rebuild there anyway. It normally takes us 4 hours to Augusta, but it took us 12 hours that day, even though we avoided the big main roads. They were already jammed with people coming up from Florida. We stayed with friends in Augusta and planned to drive on the following morning, when that area was to be evacuated, but then the storm turned again."

While Florida and Georgia breathed a sigh of relief, Floyd inundated North Carolina with several feet of rain. Giant hog farm manure lagoons built along the Neuse and Tar rivers collapsed and overflowed into waters that also filled with hundreds of thousands of dead hogs and chickens, propane tanks, gasoline and chemical slicks, car batteries from junkyards, and floating coffins from supersaturated graveyards.

I asked Frank Lapore about people who might choose to remain behind in coastal high-rises and ride out the increasing number of hurricanes that are predicted in coming years.

"Some new technologies suggest we've been underestimating hurricane wind speeds by 20 percent," he tells me. "So now we're thinking vertical evacuation for high-rises may have some problems. You run them up the first few floors from the storm surge but then you put them into higher wind speeds. So now we're saying don't take them above the fifth floor. Just look at what happened to Burger King World Headquarters [in Homestead, Florida, during 1992's Hurricane Andrew]. The storm surge took out the first two floors and the wind just took out everything above that. All that glass just got ground up like a big mix-master."

"So in terms of climate change's growing impact—"

"Ask our Virginia Key lab. Our job [at NHC] is to provide 72-hour forecasting."

If you're going to be a meteorologist, you ought to have a name like Dr. Chris Landsea. Wearing a healthy Florida tan (yes, I know), shorts, sandals, and a faded aloha shirt, this 30-something meteorologist works at the Atlantic Oceanographic and Meteorological Lab on Virginia Key, just past the Seaquarium on the causeway leading to Key Biscayne. His cluttered office includes an aquarium full of fish from the blackwater lagoon that the lab is built over and a brightly colored toy parrot. Something of a rare bird himself, being a native Floridian, Chris received high school class credits for working at the lab. When he had finished his assignments, he would go wind surfing on Hobie Beach just down the causeway. After getting his meteorology degree at Colorado State, he came back to Miami to work full-time for NOAA. His job description includes riding into hurricanes every fall aboard Gulfstream jets, P-3 Orions, and Air Force C-130s, the famous Vomit Comets that take internal readings of approaching hurricanes and are based out of Fort McDill in Tampa. "I have a weak stomach so I wear [antinausea] patches," he tells me.

"The last four years have been the busiest on record for tropical hurricanes in the Atlantic," he goes on. "The fact they didn't make U.S. landfall is more luck than anything."

I ask him if he thinks this is related to climate change.

"I was on a panel last year, an IPCC follow-up, and we found areas of hurricane activity won't change, but a French team came out proposing a small change overall in frequency, increased frequency of hurricanes. I'd go along with that."

"And intensity?" I ask.

"We see the maximum intensity of wind speed might go up 5 to 10 percent. That also agrees with recently published work out of Princeton University. So the intensity will increase for the strongest storms."

If 5 to 10 percent does not sound like a lot, consider what Dr. Steven Leatherman, a leading coastal scientist from Florida International University, has to say. Recalling that in 1992 Hurricane Andrew caused more than $25 billion worth of damage, he suggests that "if you had increased that storm by 5 or 10 percent in terms of its sustained wind speeds, those numbers would have probably popped up into the $90 billion to $100 billion category."

But there are also feedbacks within the ocean that may help mitigate those effects, according to Chris Landsea. "The current mixing that takes place in big storms brings upwellings of cooler water that weaken storms after moist, warm, ocean-generated air has started them up, so, in a sense, the ocean has a foot on the accelerator with moist warm air and a foot on the brake with cool water upwellings."

"And what role will rising sea levels play?"

"Sea surge from storms will affect Miami. Most of this county is below 10-foot elevation. So, if there's a meter sea-level rise by 2100," he pauses, "we should really begin to assess the risk of living here. I gave a talk to the Miami–Dade County South Chamber of Commerce yesterday. They're a little skeptical but after Hurricane Andrew they're much more willing to take a look at all this. Personally, I love living here and I'll take the risk, but where I live in Pinecrest we're 15 feet above sea level. That's nosebleed country for Florida. Plus I have hurricane shutters. I'm prepared. But I don't think people in Iowa should have to pay for our property here." By 2000 there was $217 billion of federal flood insurance coverage for Florida real estate that the private insurance industry refused to handle.

Despite the risks, the Florida Homebuilders Association lobbied the Tallahassee State House to limit state building codes that would require hurricane shutters and reinforced roofs on all new buildings, even as Hurricane Floyd was bearing down on the state.

Chris explains how, even in the absence of rapid climate change, local storms come in 25- to 40-year cycles linked to a 1-degree Fahrenheit warming of the North Atlantic.

"I think we're in the fifth year of a change toward these more intense storms," he tells me. "We saw this in the 1940s to 1960s, when we averaged three major storms per year. Between the 1970s and early 1990s [during the largest home construction boom in Florida history], the average dropped to

1.5 storms." He brings an ocean map up on his computer screen. Red splotches appear and disappear as he runs his program through a time line. "We've got good sea surface temperature and storm records going back to the 1870s, and based on this 120-year record, it looks like things will be getting very active over the next 25 years."

Just to add a little complexity to the story, he tells me that El Niño events, which are expected to increase in frequency and intensity with global warming, actually cut down Atlantic hurricane activity. "It [El Niño] creates vertical shear over the Atlantic in the upper troposphere that makes it harder for hurricanes to organize themselves. In 1997, for example, all these other factors were overwhelmed by the El Niño effect."

"So when we get battered by El Niño in California, that's good for Florida?"

"Better," he agrees with a grin.

Following Hurricane Andrew, private insurance companies paid out $17 billion (out of $25 billion) in storm damage. In the 1980s all U.S. weather-related claims totaled $17 billion. In the 1990s that figure jumped to more than $60 billion. This is one reason the insurance industry has begun to break with big oil over the speed with which the world needs to transition from fossil fuels to a new energy regime. During recent climate treaty meetings in Kyoto, Buenos Aires, and Bonn, a coalition including Greenpeace, the insurance industry, and small island nations emerged to counter the go-slow approach of fossil-fuel lobbyists and the OPEC nations.

At present the United States experiences around $4 billion a year in hurricane-related damages (Floyd upped the price to $6 billion in 1999). Within a $3 trillion economy, that is not considered a great deal, but what happens when the United States begins experiencing hurricane seasons with multiple impacts from what emergency planners call "$100 billion city-busters" in places like Miami, Tampa, San Juan, or below-sea-level New Orleans?

Despite grim forecasts for increased storms and hurricanes in the coming decades, the U.S. Coast Guard, responsible for search and rescue at sea during coastal disasters, is unprepared for the additional workload. Between 1999 and 2000 its budget declined from $4.4 to $4.2 billion and has been generally flatlined for years (except for congressional funds that are earmarked exclusively for drug interdiction). With an aging fleet of cutters and aircraft and too few personnel, the service commandant Admiral James Loy worries that "lack of readiness may already be costing us lives."

After my visit to the Hurricane Center, I check out South Miami Beach with Orlando, a young British visitor spending the summer apprenticing in video production with Miami friends Carl and Kathy Hersh. Carl and Kathy lost their trees and ground cover to Hurricane Andrew but suffered little structural damage to their home.

"Because most of the private insurance companies pulled out afterward, we're now way underinsured," Kathy tells me. "That just adds to everyone's anxiety. South Florida was really traumatized by Andrew. People strip store shelves like it's Armageddon whenever a storm approaches now. . . . Andrew's winds were just devastating. If there'd also been a water surge, many more people would have died. But the winds were extraordinary. They left 200,000 people homeless overnight. People lost everything—jobs, houses, their children's photographs. You also don't want to repeat what went on in the following weeks and months when everyone was looking for roofers and people were getting robbed by unscrupulous contractors. My friend had a contractor who wasn't doing anything. We went over there and videotaped him hanging out. Then he skipped town.

"This is how folklore forms, Noah's Flood kind of lore. It still marks time for a lot of people. They talk about Andrew like it just happened. It also creates stress. Child abuse, spousal abuse, divorce all increase in the wake of disasters like [Hurricanes] Hugo and Andrew.

"The devastation and smell of tar and burning waste piles in the air, that acrid smoky burning that hung on for months also reminded me of a war zone." For Kathy, who I met with Carl while covering the bloody Nicaraguan revolution of 1979, this is more than some flip media image.

"The smells and the smudged sky gave that feeling, and the big military helicopters flying over our house all day long up and down Route One." She shakes her head. "There were National Guard troops directing traffic at our corner and no one could move south without proper identification. There was a breakdown of law and order, and vigilantes were doing armed patrols to our south around Homestead. Now whenever hurricane season comes you feel like a sitting duck. We certainly did with Floyd. Friends and family were calling from Chicago, from Indiana, from everywhere, terrified, wanting to know if we were going to be obliterated. And the TV people, the forecasters, have to play a fine line, not wanting people to panic but wanting them to be ready to evacuate if told to, to take these situations very seriously."

Orlando and I saunter past South Beach's neon-bright clubs and restaurants: the Casablanca, Majestic, A Fish Called Avalon, Scandals, Rendezvous on the Beach, and Wet Willie's. As Ocean Avenue begins to crowd up we sit and eat and watch the passing parade: young women in tank tops and platform sandals, guys with pet pythons, open silk shirts, sleeveless tees. Classic cars cruise by, low-rider vans, neon-lit motorcycles attempting to compete with the neon-lit art deco buildings. Just yards across Flamingo Park and a low-lying wall is the wide sandy beachfront, compliments of the Army Corps of Engineers and their contract dredgers. We get some frozen fruit bars. The Breakwater club is jamming, its street-side stage featuring live music by Luis Menia and a flamenco dancer who moves like quicksilver in a glass. We walk past Gianni Versace's mansion, where tourists have stopped to take pictures of the steps where he was murdered.

What transformed South Beach from a low-income Jewish retirement community into a gay mecca and now a vibrant, somewhat twisted social scene is its fabulous architecture, the classic art deco apartments built following the 1926 hurricane that leveled the city of Miami and its artificial barrier island. Thousands were left homeless and 114 were killed. It is estimated that a similar storm, striking today, would cause $80 billion in damages. Nine years later, in 1935, a hurricane hit the Florida Keys, killing 427 people. At the time only 3000 people lived between Miami and Key West, about one one-thousandth of today's population. Many of those early frontier types were swamp rats, settling around what former Governor Napoleon Bonaparte Broward labeled "that abominable, pestilence-ridden swamp"—the Everglades.

A press conference is held at the Willard Hotel in Washington, D.C., to promote an $8 billion, 20-year state and federal plan to restore the Everglades. The idea is to redirect freshwater, long diverted by the Army Corps of Engineers for flood control and agriculture, back into the "river of grass." Among those in attendance are Senator Bob Graham of Florida (who leaves early), Allison Deflour, Governor Jeb Bush's policy coordinator for the Everglades, and various environmental representatives. After initial statements, they open the floor to questions.

"With sea-level rise projected at up to 3 feet, what makes you think the Everglades won't become a saltwater estuary in 50 years?" I ask.

"I think I'll take a pass on that one," Deflour smiles turning toward the enviros. They all express concern. "You look at South Florida. It's one of the most vulnerable systems in terms of storms and add sea-level rise and our friends, the mosquitoes. . . . We have to respond to this climate change issue because we're most vulnerable," says David Guggenheim from the Everglades Coalition. Ron Tipton of the World Wildlife Fund takes a different tack, arguing that "a restored Everglades will withstand impacts from this more effectively than in its present degraded state."

"It's odd," Mary Munson, cochair of the Everglades Coalition, tells me after the press conference. "In two years of discussions with 42 environmental groups, sea-level rise has never come up."

The environmentalists are not alone in failing to address the issue. The Army Corps of Engineers, assigned the lead role in restoring the Everglades water flow that it originally diverted, has not bothered to incorporate sea-level rise into its master plan schematics.

About a month later, I drop by the Everglades National Park's head-quarters and its new Visitor Center in the sultry fry heat of July. The center is a nice, low-slung shady building sitting by a swampy pond. The previous center was destroyed by Andrew. The 1.5-million-acre Everglades National Park is only about half the size of the historic Pa-hay-okee river of grass. Many of the Everglades' old marl prairies are now Dade and Broward Counties residential developments, home to some of south Florida's 6.5 million residents. Coastal areas like Palm Beach were also flooded every year, and not just by millionaires jetting in for the season. All these drained and built-up areas, plus the state's freshwater aquifer, remain vulnerable to sea-level rise and saltwater intrusion.

"The highest point in the park is only 8 feet above sea level," Ranger Rick Cook tells me, spreading a map in front of us. "A foot or two rise would effect almost everything in green," which appears to be most of the park map, not counting Florida Bay.

Of course, there are areas at greater risk. On the approach into New Orleans Airport you fly over Lake Pontchartrain and its levees, and there on the other side, down below the level of the lake's brown water, you can see curving tree-lined suburban streets, homes, lawns, kids on bikes, and parked cars. Atlantis Gardens would be a good name for this development.

New Orleans itself is 15 feet below sea level and surrounded by protec-tive levees. From parts of the French Quarter you can look up and see freighters sailing past at high tide. Only the poor are buried beneath the damp ground here where they might rise again. Elevated mausoleums keep the city's rich dry as bone. New Orleans also has a system of massive pump stations that have been operating continuously, 24 hours a day, for the past 100 years, pumping five million gallons of subsurface water back into the lake. In 1718 the early French settlers called this area the Isle of Orleans and were immediately hit by a hurricane. Today, "the town that care forgot" and the 1.3 million people living in and around it are considered a worst-case scenario for a hurricane strike by the Federal Emergency Management Agency (FEMA).

The city is still buffered by 140 miles of marshland between its levees and the Gulf of Mexico. Every few miles of that marsh is said to absorb a foot of storm surge. But with the present rate of land loss, New Orleans is expected to be a coastal city by 2040.

Johnny Glover, a fishing guide and resort owner down on the coast, has seen the change, watching land eaten away around Terrebonne Bay. "Between subsidence and sea water rising all around the world, I don't keep too much on the ground that can't be moved," he tells me. "September a year ago we had twenty days with water over my land. Our parking lot was underwater for all but maybe five days that month."

In 1990 Democratic Senator John Breaux sponsored the Federal Coastal Wetlands Planning, Protection, and Restoration Act (known as the Breaux Act). Since then some $250 million has been spent on projects expected to prevent up to 13 percent of this land loss, through restoration of the Mississippi's natural sediment transport. In 1998 a more comprehensive effort was proposed under the rubric Coast 2050, which calls for a $14 billion local, state, and federal restoration effort over the next several decades.

"We know Louisiana is experiencing sea-level rise, but it still has the capacity to create new land where the delta's natural systems are allowed to operate intact," explains Mark Davis, executive director of the Coalition to Restore Coastal Louisiana, a citizen activist group that played a key role in promoting Coast 2050.

Davis and I meet in a surfers' bar near the French Quarter that, with its quiver of boards hanging on the wall and big wave videos playing on the TV monitors, strikes me as weirdly out of place, like a tiki bar in Omaha.

"The Army Corps has given us a levee system that's traded periodic Mississippi river flooding for permanent coastal flooding and that to me is a bad deal," Davis says. "We built highways to move troops coast to coast and built a huge military machine on the chance that the Russians might threaten us. We buy insurance out of fear of injury or disease, not because we know it's going to happen. It's the same thing here. We have to act because the costs if we get it wrong are too high."

Does that mean he believes he can mobilize enough popular will to turn around the forces of sprawl and development, not to mention the Army Corps, oil drilling, and canalization that contribute to subsidence and land loss?

"If our advocacy is inadequate to the task," he warns, "then a hurricane will make the case for us."

On the West Coast, rising seas is only one of the climate factors that have begun to alter the marine ecosystem. The 1999 report "Confronting Climate Change in California," put out by the Union of Concerned Scientists and the Ecological Society of America, points to three areas of risk: increased frequency and severity of El Niño storms, loss of fisheries from a warming Pacific, and loss of wetlands and hardening of the coast in response to sea-level rise.

Projected increases in the number and intensity of the El Niño Southern Oscillation (ENSO) are based on computer modeling and the dozen El Niño events—involving warming Pacific sea surface temperatures—that were recorded in the twentieth century. These include two severe El Niño events in 1982–1983 and 1997–1998. Along with coastal storms and increased runoff from flooding, the El Niño effect has been marked by dramatic shifts in offshore biology. During the last El Niño, fishermen in San Diego were catching tropical tuna while barracuda were being landed off San Francisco.

Examining long-term trends recorded by the California Cooperative Oceanic Fisheries Investigations (CalCOFI), scientists identified a gradual warming of sea surface temperatures that has led to an increase in southern species of kelp forest fish (goldfish-like garibaldi) at the expense of northern rockfish (an edible commercial species). They also noted a northern expansion of sardine populations (whose 1947 collapse led to the creation of CalCOFI). More disturbing, the 1970s saw an abrupt jump in water temperatures off California that has persisted. According to the report, "The warmer temperatures of the past two decades have been accompanied by reduced mixing in the water column, reduced upwelling of nutrients, and widespread declines in algal productivity along the California coast. The decline in productivity has been followed by equally large changes . . . including declines of sea birds and an accelerated decline in yields of commercially fished species."

A projected sea-level rise of 1 foot in the San Francisco Bay, Sacramento Delta area, and increased storm surges (which have been recorded since 1970) are expected to turn 100-year flood events into 10-year floods. One predictable response to this increased flooding activity will be increased demand by coastal property owners and delta farmers for more levees and protective seawalls that, while providing temporary relief, will speed up erosion and prevent wetlands and other natural systems from adapting to the coming change.

On the Atlantic's Chesapeake Bay, Maryland's Smith Island, a small crab-fishing community of 450 people, has shrunk from 10 by 6 miles to 8 by 4 and continues to lose some 30 feet of shoreline a year. State policy advisers point to Smith Island as an example of how sea-level rise has to be taken into account in regional planning for bay restoration and cleanup. But now that one-third of Maryland's bayfront property has begun to collapse, including the lawns of middle-class neighborhoods in populous Anne Arundel County, state legislators have decided to act defensively. They have set up a special task force to reexamine anti-erosion efforts, including the building of seawalls, breakwaters, and other stone and concrete barriers that were abandoned years ago.

In trying to stem the tide by armoring the coasts, the legislators would be choosing a scientifically discredited but widely popular approach. Even as the risks from hurricanes and rising seas increase, more and more Americans are moving to the beach. Developers, ever attuned to market trends, are building them ever grander castles made of wood and stucco. And all this high-risk behavior is being supported and encouraged by a government that hasn't learned how to just say no.

Chapter Seven

Paradise with an Ocean View

Where Margaritaville meets Nantucket—
A 54,000 sq. ft. club & spa.
A planned golf clubhouse and 54 holes of golf.
A yacht club with a deep-water marina.
And homes with golf, nature or water views.

         —REAL ESTATE AD, NEW YORK TIMES

What the sea wants, the sea will have.

         —A SAILOR'S SAYING

Barrier islands are like geology on amphetamines. Unarmored, they tend to move, by the decade, year, season, or sometimes in a single stormy day. It's a natural process that can be strikingly beautiful, even awe inspiring, provided you haven't just closed on a $2.7 million beachfront dream home. Unfortunately, more and more wealthy Americans are doing just that, moving to or buying second homes in places like Fire Island, Easthampton, Hilton Head, Sea Island, Ocean Reef, or Captiva. Other folks are buying high-rise condos in Ocean City, Myrtle Beach, Gulfport, Coronado, and Honolulu or flocking like lemmings to new housing developments built on filled-in marshes and flood plains all along the coastline. And while lemmings don't actually jump off cliffs and drown themselves in the sea, if they did, there would undoubtedly be a number of government programs offering them flight insurance and full-coverage for any water damage.

More than 141 million people, about 53 percent of the U.S. population, now live within 50 miles of a marine coastline, with about a million new arrivals showing up each year. Every day some 2000 new homes are built along our coasts. Fourteen of America's 20 largest cities are coastal, as are 17 of America's 20 fastest-growing counties. Every week 3400 new people arrive in southern California and 4400 in Florida. The coasts (excluding Alaska) are now three times as densely populated as the rest of the country. And while various government reports have expressed concern about associated problems of erosion, sprawl, marine pollution, loss of wildlife habitat, and hurricane safety, federal programs continue to encourage people to move to the beach.

"You might need somethin' to hold on to, when all the answers they don't amount to much." Bruce Springsteen sings on the car stereo as I drive past cherry trees in bloom, old clapboard houses, and corner groceries, across railroad tracks, and down to the beach in Asbury Park, New Jersey. This is not one of those recently restored maritime waterfronts full of colorful seafood restaurants and potpourri-scented lighthouse boutiques. With its eroded tax base and absentee landlords, it looks more like the beachfront in wartime Beirut, dominated by abandoned dance halls and a skeletal, unfinished high-rise. There are weedy lots, boarded-up stores, and street lights left on in the middle of the day. I climb a fence and go out on the warped boardwalk, beyond which rock jetties jut out into the sea. A guy in a wheelchair and army pea coat is the only other person in sight. I climb over a wooden rail and make my way across the sand. Gulls are flying about looking for a handout or something newly dead. There's a "No Swimming" sign on the beach, which is being scalloped out between the jetties. For years people built rock groins and jetties like these to retain sand in front of their homes and hotels, while eroding the beaches downdrift.

Those people would then build their own jetties to capture waterborne sand carried in the nearshore current, until eventually there was an ugly series of rock piles ending in a stretch of severely eroded shoreline. At that point seawalls were often erected, only to collapse in two, 10, or perhaps 20 years as increased wave energy generated by their steep hard surfaces undermined the walls' own foundations and drove any remaining sand back out to sea. There's a term coastal geologists use for this process of hardening the shore. It's called "newjerseyization."

Driving through Asbury Park I notice the Albion Motel and Rainbow Room are closed as is the Stone Pony, where the Boss used to play with his band. Heading north along Ocean Avenue, the scene quickly shifts to one of mansions, private beach clubs, and condos and the ocean is lost to sight behind them. When the view reappears in Long Branch, it is of a large seawall.

On the mainland side of the road are the modest homes, shops, streetside bars with Bud Lite banners, sandy sidewalks, and chain convenience stores that have become iconic of many American beach towns or, in this case, wall towns.

In Monmouth Beach, the next town up the "shore," I park at a bus stop and climb the wooden stairs next to the shelter that's marked "Public," the first public access sign I've seen in five miles. Unfortunately, all the side streets have no parking signs that warn if you are not a resident your car will be towed. The cobble rock and cement wall is about 15 feet high and 12 feet across the top. There are numerous other stairs up and down this taxpayer-built wall, but they're all marked "Private—Keep Out." On the other side of the wall is a bermlike dune and wide beach recently pumped onto shore by the Army Corps of Engineers. I'd like to check it out but am afraid a cop might come along and ticket my illegally parked car. If there's ever been a less user-friendly beach I can't imagine it, except perhaps Normandy on D-Day.

I put up that evening at the MiraSol Motel on the 150-yard-wide coastal strip of Sea Bright between the Shrewsbury River and a newly created beach that I can see over the wall from my second-floor room.

Sea Bright, some 25 miles south of New York City, is the starting point for the largest and costliest "beach nourishment" project in the United States. It is an Army Corps of Engineers operation (65 percent federal money, 35 percent state funds) designed to build up beaches along the entire 127-mile length of the Jersey Shore. Begun in 1994, final costs are projected to run as high as $9 billion, or more than $60 million a mile. Similar projects are underway in South Florida, Mississippi, and Texas.

"You can't believe the pressure we get from the states for beach replenishment. It's a huge, insatiable demand," says Dr. Joseph W. Westphal, assistant secretary of the army for civil works, the political appointee who oversees the Corps of Engineers.

Even with 150 feet of beach expected to disappear for every foot of sea-level rise, Sea Bright's new strand could last for decades, provided there are no major storms. But storms have already washed away Monmouth's new beach three times in the past four years, making it what corps engineers call an erosion "hot spot."

Sandy Hook, the last curling spit of sand just north of Sea Bright, is covered in dry grass, cattails, scrub oak, and pine. I follow directions to its old Spanish-American army fort that, like San Francisco's Presidio, is one of America's glories, a coastal park. This one comes complete with a Coast Guard station, a national marine fisheries lab, a flock of Canada geese, and, along Captain's Row, the offices of the Littoral Society, a nonprofit activist group with some 6000 members, mainly in the New York–New Jersey and Delaware area. I enter the 100-year-old officer's quarters by the kitchen door, where I encounter the society's director, Dery Bennett, steaming clams on an old four-burner stove. At 68, Dery stands 6 feet, 3 inches, more gristle than fat with skin like salted salmon leather, curly gray hair, and brown eyes that can go from mournful to mirthful, depending on the light. Taking the clams from the pot and placing them in a china bowl of hot water, he cracks them open and starts to eat.

"I live in Fairhaven. Moved there in 1968. They closed the clam beds 32 years ago and reopened them last year, so this is the first time I've gotten to try these," he says while chewing and crunching.

"How is it?"

"A little overcooked and sandy." He dips another clam in the hot water. "Not bad. Things have improved. They got the owners to compost the manure at Monmouth Park Racetrack. Reduced the farm runoff. I just raked these on the riverbed yesterday, the second-to-last day of the season. When they opened the beds last year, guys were in there taking six to eight bushels a day, that's six to eight hundred dollars. They rebuilt that fishery over 30 years but they've hammered it back down in a hurry."

He finishes his clams and leads me up to his cluttered office. The whole building is cluttered but cool, filled with stuffed fish—wahoos, rays, striped bass, bluefish, a swordfish with a Santa cap, a big hammerhead. In his office Dery shows me a map of dredged channels, drowned rivers behind New Jersey seawalls built in the 1930s and 1940s.

He begins talking about the corps beach project, although I can't quite take my eyes off the goosefish skull hanging on his lamp like a wide-brimmed hat. "They plan for 50 years protection," he tells me. "Earlier seawalls were put up to protect towns like Sea Bright, and the direct result was the loss of the beach. The corps pumping and then maintaining the sand. It's supposed to last seven years. So they'll redo it every seven years."

The day I met with Dery, Congress passed the $6 billion Water Resources Development Act (WRDA), the annual funding mechanism for dredging harbors, building levees, and pumping sand.

Dery is not a fan of pumping sand.

"You've got a tough fight if what you're saying is don't save the beaches," I point out.

"I think if there's winners and losers, I'm on the side that's losing right now," he admits. "New Jersey just voted up to $25 million a year for beach sand pumping, but not a nickel for improved public beach access. Of course most of the money still comes from the feds. That's why the states like it, cause of the pork, the salt pork. It's welfare for the rich and the damp."

"To say this only benefits the rich is disingenuous. There's no question all of society in coastal states benefit from opportunities for recreation," counters Jim Saxton, the environmentally concerned Republican Congressman from New Jersey's Third District, which includes much of the shore. "Economists have told me that tourism accounts for 50 percent of New Jersey's gross domestic product. So it's in our interest to make an investment in beach nourishment and replenishment. We have families that save to rent a cabin or condo for a week or two during the summer. You'll see hundreds of them in a single block along the shore. We also have day trippers who'll come to the beach for a day in the summer. I just think it's short-sighted to say this only benefits a few big home owners."

Across the New York bight from Sea Bright, on Long Island's south shore, sand replenishment has clearly benefited a few big home owners, along with some unscrupulous real-estate developers. Following a 1962 Ash Wednesday storm that destroyed a large number of homes, the Army Corps of Engineers New York District began a 15-year project to replenish the sand at Westhampton, building 21 groins, 480-foot-long rock spurs, running perpendicular to the beach. Since the nearshore sand transport ran east to west, the way to minimize the erosion effect these structures would inevitably create would have been to build them west to east, which is what the corps planned to do. But, according to a 1983 *New York Times* account, wealthier property owners on the east end "used their political influence . . . and the county told the Corps of Engineers that it would not pay its local share unless the project was changed to go from East to West." That change assured the richest homes would be the first to get storm protection, even at the cost of starving the residential beaches to their west.

As a result, during a series of storms in the early 1990s a new inlet formed across the erosion-narrowed western end of the barrier island at Dune Road. Two hundred and fifty homes were cut off from the rest of the island, but only 90 of them were still standing after the storms had passed. In order to strengthen a class-action suit against the county, state, and federal authorities, 150 property owners from the area incorporated as the Village of West Hampton Dunes. In 1994 they won a legal settlement in which the federal government, along with the state and county, agreed to fill in the newly formed channel and build a 400-foot-wide beach along the vulnerable Dune Road at an estimated cost to taxpayers of $75 million.

In the five-part series "Shoreline in Peril," published in *Newsday* in August 1998, reporters Thomas Maier and John Riley showed how the mayor and other West Hampton Dunes officials quietly bought up a number of the town's sunken lots from more desperate property owners, roughly tripling their value once the government sand was in place and new construction started on vacation homes in the half-million to three-quarter-million-dollar range (that also qualified for cheap federal flood insurance).

"What we're doing here is permitted by law and is moral and legal," claimed the newly prosperous Mayor Gary Vegliante, who went on to suggest that, "What happened in West Hampton Dunes is a model for the nation on coastal policy."

An alternative model exists along 1500 miles of Massachusetts' ocean beaches and bays where, since 1978, the state has banned construction of seawalls and jetties. "We lost half our wetlands up 'til '78 and have lost very little since," says Jim O'Connell, the former state coastal geologist who now directs the Woods Hole Coastal Research Center. I'm sharing a bench with him along the Shining Sea Bike Path overlooking Nantucket Sound on Cape Cod.

"Natural erosion creates bays, estuaries, and barrier beaches," he explains, "But if you armor your uplands you lose your beaches and wetlands. If you put seawalls up along the Cape Cod National Seashore it would disappear in less than a century." An older white-haired woman in pink cotton pants and a striped shirt stops on the path in front of us. "I love to see people sitting there. My husband and I, this was our favorite spot to sit here and watch the ocean." She resumes her power walk. I notice a bronze plaque on the bench back: "In loving memory of Bill Port, 1919–1997. Shirley."

"Thank you, Shirley!" I call out. She turns her head, smiles, and continues on her way.

I'm swimming in warm, milky-white waters just off Miami Beach. The water's cloudiness comes from suspended silicate not yet settled out of the sand slurry that's being pumped a few hundred yards south at Government Cut. I scoop some sand off the bottom and find it grainy and thick with shell. Offshore, there is a big four-legged dredge platform owned by the Great Lakes Dredge and Dock Company and a large sand pipe coming ashore at a fenced-off area near the shipping channel where jet skis and cigarette boats are buzzing around and young Haitian teens are jumping off a high wooden fishing pier. The sand slurry (about 85 percent water and 15 percent sand) has stopped pumping for the day and a Caterpillar dozer is spreading the sand in a rampart down along the water's edge.

This 13-mile stretch of beach, part of 150 miles of Army Corps of Engineers beach projects in Florida, is the nation's second largest sand nourishment program after that of New Jersey. Miami Beach once had a natural strand but seawalls built by hotel owners eroded that away, and now the corps is moving sand from offshore to onshore to maintain this new beach. There is only one problem.

"We're exhausting the last offshore sand in Dade county," says David Schmidt, chief of the Coastal Navigation Section for the corps Jacksonville District. "So we're looking to aragonite [sand] either upland or barged in from a foreign country. In the Bahamas and the Turks and Caicos, aragonite precipitates out of the water. But the Bahamas won't give us any because they're in competition with Miami and want tourists to come to their nice white beaches. The Turks is willing, but the American dredging industry doesn't like the idea of foreign sand. They think the work will go to foreign companies. So they got Congress to pass a law saying foreign sand can't be used where there's American sand available."

It sounds like protectionism to me, although not the kind that could work for our coasts.

Along with flood control and beach replenishment the Corps of Engineers also dredges some 400 million tons of sediments a year in shipping channels, marinas, harbors, and ports, about three times the amount of earth moved to build the Panama Canal. Most of this dredge spoil is dumped into rivers, bays, and estuaries or used for landfill. Some 60 million tons goes to 100 ocean dumpsites selected by the Environmental Protection Agency (EPA).

While most shipping channels leading into America's major ports contain fairly clean muds or sand and seagrass bottoms, some inner harbors, particularly in older cities and military ports, contain sediments contaminated with high levels of mercury, heavy metals, PCBs and dioxin. For years the EPA refused to dredge at a dioxin-contaminated Superfund site off Palos Verdes, California, for fear that disturbing the bottom would stir up the contaminants. At the same time, it approved dredging and ocean dumping for the Port of Newark's inner harbor, even though testing showed the area to be highly contaminated with dioxin from an old Agent Orange factory. This led to a series of protests, lawsuits, and a 1997 decision by the state of New Jersey to ban the disposal of contaminated sediments off its shores. The EPA agreed to shut down its ocean dumpsite, which by then had grown into a subsea mountain 1.5 times the size of Manhattan.

In the fall of 1999 the Littoral Society, Clean Ocean Action, Coast Alliance, and other marine conservation groups held a Washington, D.C., Summit on Sediments and released a 152-page report called "Muddy Waters: The Toxic Wasteland Below America's Oceans." The report compared the EPA and Corps of Engineers ocean dumping of untreated sediments to "a medieval man tossing the contents of his chamber pot out the window." But rather than simply put out bumper stickers reading "Save the Mud," the coalition proposed a number of practical solutions. These were based on the idea of combining pollution prevention (not allowing any new contamination) with cleanup technologies being tried out in the New York–New Jersey Harbor area by government and commercial enterprises.

These techniques include chemical, pressure, and thermal treatments in which contaminated sediments are run through big rotary kiln incinerators or are mulched with wood chips, compost, and manure to form commercial soils that can then be cleaned by cattails and other wetland plants that absorb heavy metals. Other treatments involve the use of recoverable solvents that remove pollutants from mud or the locking of contaminated sediment slurries into cement or glass-like materials useful for aggregate, landfill, and roadfill.

Oddly, the coalition's proposal that would seem the easiest to carry out, the idea that ports not do any unnecessary dredging, may prove to be the most difficult. This is because of the increasingly unbalanced relationship between America's ports and the globalized shipping industry.

Today, more than 97 percent of America's non-NAFTA trade goods, about a billion metric tons a year, are carried by ships across the Blue Frontier. And while West Coast trade is fairly concentrated in three major port areas—Seattle–Tacoma, Oakland, and America's largest port, Los Angeles–Long Beach—there are literally dozens of East Coast and gulf ports being played off against one another by the world's shipping companies. Boston, New York, Philadelphia, Baltimore, Norfolk, Charleston, Savannah, Jacksonville, Miami, Tampa, New Orleans, and Galveston-Houston are among those vying to become megaports, dredging ever deeper and wider channels in the hope of accommodating a new generation of container ships called Post-PanaMax. These are freighters too large to fit through the Panama Canal (or into most existing ports).

In 1956 there was a revolution in shipping when Malcolm McLean, a North Carolina trucker turned entrepreneur, introduced "the box," a reinforced truck trailer with its wheels removed. He put 58 of these boxes on the *Ideal X*, a ship he modified and sailed from New York Harbor to Houston. That voyage marked the beginning of the end of net-loaded break-bulk cargo.

The longshoremen's union, led by San Francisco's Harry Bridges, recognized that containerization would mean fewer men working the docks. Rather than fight the technology, they negotiated contracts that allowed their older members to retire with their benefits intact. This allowed the union to maintain control of the remaining jobs (crane operators, yard truck drivers, and so on). The continuing power of the union was demonstrated during December 1999's "Battle in Seattle," when longshoremen staged a one-day shutdown of West Coast ports in solidarity with the anti–World Trade Organization protesters.

Today, containerization and intermodal transport (linking shipping docks to railroad spurs and highways) has dramatically decreased costs and increased speed of delivery of global marine commerce. Shipping costs have become so cheap that a growing number of giant gantry cranes, those super-stars of the ports, are now built in China and shipped whole from Shanghai. These cranes lift and load what are now called twenty (20)-foot-equivalent-unit (TEU) containers. Modern container ships can carry as many as 3000 TEUs, both within their hulls and stacked up to 12 high on their decks.

But even as the oil transport industry is eliminating supertankers in the wake of numerous maritime disasters, the shipping industry continues to expand the size of its container ships. In 1998 Denmark's Maersk Shipping sent its aircraft carrier–sized 6000 TEU ship *Regina Maersk* on an East Coast tour of the United States. It first had to stop at the natural deep-water port of Halifax, Canada, to offload part of its cargo so it could ride high enough to fit through New York's 40-foot-deep shipping channel and dock at Port Newark. Still, shipping companies are planning even larger 8000 and possibly 10,000 TEU ships that will require even deeper ports.

An alliance of Maersk and Sea-Land, whose leases with New York were scheduled to end in 1999, offered to move their business (350,000 TEUs a year) to whatever port offered the best incentives, including dredging to 53-foot depths.

The ports could have refused to accept these hard-to-handle megaships or designate a single deep-water hub and series of feeder ports. Instead, they reverted to form, trying to steal each other's business. The Port of Baltimore and State of Maryland put together the largest financial incentive in the state's history to lure away New York's leases. "The thing is we already have a 50-foot channel we've dredged and don't think that in 10 years they [the Port of New York–New Jersey] will be able to get down that deep," argues Frank Hamons, development manager for the Port of Baltimore. "They have a lot of rock to blast, whereas we have a soft bottom."

In the end, the shipping companies opted to stick with the Big Apple, having squeezed as much juice out of it as they could. In the August 1999 issue of *Container Management* magazine, Lillian Borrone of the Port Authority of New York and New Jersey bemoaned the fact that as the ship-ping industry consolidates and sets the terms for doing business, ports, which are mostly public enterprises, continue to favor cutthroat competi-tion over alliance-building that could improve their bargaining position. One result is $7 billion in new deep-channel dredging and expansion pro-jects set through 2004.

Many of these dredging projects are hard to justify either from an envi-ronmental or an economic point of view. Attempts by the Georgia Ports

Authority to deepen the river channel into the Port of Savannah, for example, has generated opposition from the city of Savannah, the local business
community, and conservationists. According to hydrologists, deepening
the channel will allow a saltwater ridge to flow upriver, which could kill up
to 10,000 acres of the 27,000-acre Savannah National Wildlife Refuge
located just across from the town's riverfront tourist district. Businesses
that withdraw water from the river are also worried about silt and salt intrusion, as is the city, which fears it might have to relocate its water treatment
plant. Still, the 1999 WRDA authorized $230 million to deepen the
Savannah harbor from 42 to 48 feet.

When the U.S. Fish and Wildlife Service, which runs the wildlife
refuge, objected to the dredging, Republican Representative Jack
Kingston, a dredging supporter, went ballistic. A leaked Georgia Ports
Authority note reads: "JK [Jack Kingston] needs us to kick F&W's ass in
the paper. . . . Don't let rinky-dink agency beat us." Still, economic projec-
tions for the port show that its dredging and expansion will only work if
Savannah can steal business away from the ports of Brunswick, Georgia,
Charleston, South Carolina, and Jacksonville, Florida, which are also
expanding.

Ironically, along with dredging ports, the Corps of Engineers and EPA
are responsible for protecting wetlands and approving permits that allow
wetlands to be filled in or altered. More than half of America's coastal wetlands have been lost to development over the life of the Republic, including
95 percent of California's, with the greatest losses occurring in the 1970s
and 1980s. Since President George Bush's pledge of "no net loss of wetlands" back in 1989, the rate of loss seems to have slowed but not stopped.
It's hard to tell because the corps and EPA have failed to generate data that
can definitively say either way.

For generations, wetlands were perceived as dank and dangerous
swamps, home to snakes, alligators, and the occasional philosophical possum. Since World War II some 50 million acres of farmland have been
paved over by urban and suburban development, while some 53 million
acres of wetlands have been filled in for agricultural use. This was considered a reasonable trade-off until science began identifying wetlands as a key
habitat for migratory birds and wildlife, a nursery for 75 percent of our
marine fisheries, a protective barrier for coastal storms and hurricanes, a
filtration system for pollution, and a natural recharger of freshwater
aquifers.

Coastal wetlands and estuaries often trail off into sea grass meadows
that also secure the nearshore bottom, reduce turbidity, and provide vital
habitat for juvenile fish and shellfish. But significant losses of seagrass have

occurred as a result of sediment and nutrient runoff, vessel wakes, and scarring from dredges, anchors, and shallow-draft vessels including personal watercraft.

Visiting the Georgia coast south of Savannah, I'm overwhelmed by the soft beauty of its dank charms. About one-third of the entire eastern seaboard's salt marshes are located along this narrow 120-mile coastline. For a fellow from California, where a reedy duck pond near the ocean is considered a maritime treasure, seeing extensive hammocks of live oak and Spanish moss hanging down over the banks of meandering coastal river marshes that flow as far as blue herons can fly, is just too nice. The locals attribute their coast's relatively pristine condition to 8-foot tides (that make building difficult), wealthy carpetbaggers (who bought up barrier islands as family retreats), tree farms (that dominate the local economy), and insects (that bite).

They are also thankful to Savannah's Jim Williams, the accused murderer in the best-selling book *Midnight in the Garden of Good and Evil*. It was his sale of Little Tybee Island to Kerr-McGee for a planned phosphate mine back in the 1960s that outraged and inspired Georgia to establish one of the strongest coastal wetlands protection laws in the United States.

Still, even here people worry that their natural coast may not stay this way for long. "If we don't do the right things now we'll be like Florida," warns Robert DeWitt a blue crab wholesaler in Darien. "Right now we don't have a coastal population, we've got sand gnats."

The same cannot be said for urban areas like New York and Los Angeles, where people outnumber sand gnats, and the last remaining coastal wetlands are highly valued, by real-estate developers.

The New Jersey Meadowlands, just across the Hudson River from New York, is an urban-wetland interface that only a child of the city (or juvenile fish) could love. In his book *Meadowlands*, author Robert Sullivan calls it "a thirty-two-square-mile wilderness, part natural, part industrial, that is five miles from the Empire State Building and a little bit bigger than Manhattan . . . a good place to see a black-crowned night heron or a pied-billed grebe or eighteen species of ladybugs, even if some of the waters these creatures fly over can oftentimes be the color of antifreeze."

It is here amidst numerous old towns, factories, garbage mountains, the New Jersey Turnpike, and 8000 remaining acres of marshland that the Mills Corporation of Arlington, Virginia, and its German equity partner Kan Am hope to fill in more than 200 acres of tidal wetlands. They would like to replace them with a 2.1-million-square-foot shopping mall and entertainment complex. It would be the largest mall in New Jersey, eclipsing the nearby Meadowlands mall and competing with the New Jersey Garden State mall.

According to the Clean Water Act, the Corps of Engineers can only approve destruction of wetlands along navigable waterways for "special aquatic sites," like ports, marinas, and municipal fishing piers. It is not supposed to approve any wetlands destruction unless "mitigations" (restoration or creation of new wetlands) ensure at least a doubling of acreage.

The Mills Corporation promised to improve and preserve 380 acres of adjacent "degraded wetlands" if it was granted its development permit. That sounded good enough for the corps and EPA until they ran into opposition from coastal protection and urban planning groups. These outfits, including Clean Water Action and the New York–New Jersey Baykeeper, a kind of waterborne Lone Ranger, argued that by opening some long-closed tidal gates, the corps could restore all 600 salt marsh acres at the site. They suggested incorporating this acreage into a national urban wildlife refuge, an idea for which the U.S. Fish and Wildlife Service has grown sympathetic.

By 2000 the mall plans were still up in the air, backed by the Hackensack Meadowlands Development Commission but editorially opposed by the *Bergen Record,* which suggested the Mills Corporation consider relocating its mall to a new sports complex being built in Newark's urban core for the New Jersey Nets and Devils.

The year 2000 also saw the first aboveground construction begin for a $7 billion "dream city" called Playa Vista (Spanish for "beach view"), located on the last undeveloped coastal valley in Los Angeles. The 1087-acre Ballona Valley north of Los Angeles International Airport by Lincoln Boulevard missed Los Angeles's postwar building boom only because it had already been bought up by Howard Hughes. For years, the mysterious Mr. Hughes was content to maintain a single 60-acre aircraft factory on the valley's far eastern end, even as Marina Del Rey and other developments sprang up along the edge of the valley's coastal wetlands. In 1947 the Hughes aircraft factory completed production on its most famous work, the Spruce Goose, a seaplane so large it could barely rise off the water.

Today, Wall Street investment giants Goldman Sachs and Morgan Stanley lead the Playa Capital consortium that is developing the historic valley site. But it was Hollywood's DreamWorks, which dropped out of the project in 1999, that received most of the media attention. DreamWorks principals include movie titans Steven Spielberg, David Geffen, and former President Bill Clinton's amigo Jeff Katzenberg. DreamWorks and its partners promised to spend $13 million to restore 188 acres of degraded wetlands along the shoreline and create additional wetlands and wildlife habitat around Playa Vista's pedestrian malls. During the 1996 presidential campaign Geffen hosted a $25,000-a-head party that raised $3.5 million for Bill

Clinton and Vice President Al Gore. Spielberg, Geffen, and Katzenberg also personally contributed more than $1.5 million to the Democrats between 1990 and 1997. So there may have been more than environmental policymaking going on when in the spring of 1995 Al Gore's staffers started calling up local green groups in Los Angeles. They wanted to know how the enviros would react if the vice president were to come to town and endorse Playa Vista as a model of sustainable development. The Sierra Club told Gore's office they would picket him if he showed up and the trip was canceled. Still, Playa Vista was selected by President Clinton as one of five PATH (Partnership for Advancing Technology in Housing) communities promoted as models of sustainable development, now known as "smart growth."

Opponents of Playa Vista, pointing out that Los Angeles has the lowest ratio of parks to pavement of any major city in America, have spent a decade calling for restoration of the valley as a tidal wetlands and city park. They have also complained about projected traffic and air pollution from the 200,000 additional car trips a day Playa Vista expects to generate.

Reporting on the conflict several years ago, I was impressed by the bizarrely contrasting visions that Los Angeles can conjure up for a visiting journalist.

It was raining and misty on my first day as I stood on the Westchester Bluffs looking a mile west toward Marina Del Rey and the ocean. Large sheeting pools of water in the still green valley provided easy pickings for white-feathered egrets and insect-hungry frogs, whose crickety trillings carried on the wind. Soon I was down at ground level climbing through a broken hurricane fence with activists Marcia Hanscom and Bruce Robertson. My running shoes quickly soaked through as we wandered among pickleweed, cattails, and chest-high foliage, spotting a hovering kite, the fuzzy end of a rabbit, and half a dozen great blue herons foraging along the edge of a brackish creek before lifting off in a great beating of wings, 6 feet across. I asked them (the activists, not the herons) if they thought they could win their battle to save an ocean view valley in the heart of Los Angeles. "Well, it's kind of like a Spielberg movie," Hanscom grinned, "where the little guys take on the huge corporate developer."

The next day was bright and sunny as I was driven upvalley on an old two-lane tarmac road by Doug Gardner, the friendly architect turned site manager for Playa Vista. The first animal we spotted was a large African elephant rearing up on its hind legs next to a moving van. We were at one of Howard Hughes's old aircraft hangars that had been converted to a movie studio, where they were shooting *George of the Jungle*. Inside we wandered through several acres of fake canopy-thick forest and softly yielding earth made of potting soil and wood chips. There were klieg lights and big 35-millimeter cameras on dolly mounts and caterers scuttling around with trays of straw-studded coconut shell drinks for the cast. I recognized a unique kind of dream magic taking place here: Hollywood's always seductive ability to suspend our belief, whether creating a tropical rain forest in an aircraft hangar or pitching the idea that the only way to save the last coastal wetland in Los Angeles is to build a new city on its shore.

Still, most of today's wetland losses are not big and controversial and played out in the media, but small and little noted. It's an acre here, two acres there, usually permitted by the Corps of Engineers without inspection. A whole new consulting industry has even grown up to help developers navigate government regulations and permit requirements on what they refer to as "environmentally challenged sites." By "environmentally challenged" they mean (George Orwell would appreciate this) pristine coastal wetlands and forested islands containing endangered species.

In 1999 Public Employees for Environmental Responsibility (PEER), a whistle-blowing organization made up of people working inside various government agencies, issued a report on the Corps of Engineers. Based on Freedom of Information Act requests and interviews with corps regulators around the country, it found wetlands protection enforcement was down dramatically with 40 percent fewer inspections and 80 percent fewer court cases brought against people who destroyed wetlands between 1992 and 1998.

John Studt, the corps regulatory chief, did not dispute the figures but argued that enforcement was down because compliance was up. "There are fewer violators involving substantial impacts to the aquatic environment because more people are aware of environmental laws," he claimed. To which Jeff Rouch, executive director of PEER replied, "How the hell does he know?"

I asked Joe Westphal, the assistant secretary of the army who oversees the corps, about the PEER report.

"We do plan to maintain inspection and enforcement," he insisted. "But any regulatory agency is going to have trouble keeping up and enforcement is the most difficult work. We're not a police force to be out there and investigating."

Westphal, it turned out, was also having difficulty keeping up with his own generals. In February, March, and April 2000 the *Washington Post* ran a series of articles exposing internal plans by the military commanders of the corps to increase their budget 50 percent over five years through rapid expansion of their water projects. This was to take place without apparent regard to the economic necessity or environmental impacts of the works.

"Oh, my God, my God. I have no idea what you're talking about. I can't believe this," Westphal told a *Washington Post* reporter when asked to comment on the generals' plans. The corps' military "green shirts" seemed to be in thrall to some primal spirit force beyond their control, a creature half warrior and half beaver.

On March 30, 2000, in the wake of media and congressional inquiries, the Pentagon announced a series of reforms designed to restore civilian control over the Army Corps of Engineers, then reversed itself again under pressure from Republican Senators Ted Stevens, John Warner, and Bob Smith.

Meanwhile, America's coastal wetlands continue to decline as tidal flows are cut off by roads and houses, and Spartina (sea hay) is gradually replaced by the prettier but less productive swamp grass called Phragmites ("frag").

The same process of fragmentation and loss that is happening to wetlands can also happen to coastal cultures when traditional fishing towns, beach communities, and waterfront cities fail to maintain their civic vigilance. That's when they can morph into overbuilt urban seawalls like Atlantic City or Gulfport, Mississippi, driven by casino, condo, and real-estate values that measure life on the Blue Frontier solely by the dollar value of their oceanfront footage. Unfortunately, prime beachfront real estate, like prime-rib steak, tends to be transitory in nature.

In 1968 FEMA, worried about the crowding of the coasts and disaster risks faced by new beachfront residents, came up with a plan. If homeowners met certain basic safety standards in beach construction (like putting houses on stilts), they would qualify for a newly created National Flood Insurance Program (NFIP). FEMA convinced Congress this would reduce individual risk while shifting the burden of hurricane disaster relief onto policyholders. They would guarantee a large insurance pool by making the rates so inexpensive lots of people would buy the policies. This idea worked for awhile, about as long as the historic lull in Atlantic hurricane activity of the 1970s and 1980s. As soon as Hurricanes Hugo, Andrew, George, Fran, Floyd, and their ilk started coming ashore, the program turned into a huge money loser and the largest financial exposure the federal government now faces, with over $521 billion in flood insurance policies.

But the greatest impact NFIP has had is the fueling of the coastal construction boom of the past 30 years, which makes the western land rush of the nineteenth century pale by comparison. As Cornelia Dean points out in her book *Against the Tide,* until NFIP came along it was almost impossible to insure beachfront property against flooding. As a result most bankers refused to issue mortgages. That meant beachfront home owners had to pay cash and tended not to build bigger than they could afford to lose. But once the feds started offering beachfront flood insurance (for an average $300-a-year premium), real-estate developers found mortgage bankers more than willing to lend them money, setting off a tsunami of new coastal construction. When Hurricane Frederick struck barrier islands off Alabama in 1979, for example, shattered beach cottages at Gulf Shores were quickly replaced by 14-story fully mortgaged condominiums.

Federal flood insurance covers up to $250,000 for property damage to homes and $500,000 for businesses. Storm-battered beach communities are also eligible for low-interest small business loans, federally funded reconstruction of highways, roads and bridges, and sand replenishment.

"The problem is the government won't operate like a business. It won't cancel policies or increase premiums when a house is destroyed like a real company would," complains Steve Ellis, a maritime specialist with the Washington-based Taxpayers for Common Sense. "Erosion and sea-level rise aren't even factored into their rates."

FEMA insurance also allows policyholders to make repeat claims, encouraging people to rebuild in harm's way. ABC-TV correspondent and free-market champion John Stossel aired a *20/20* piece on FEMA's insurance program after his weekend beach house in Westhampton was washed away in a storm. "It was an upsetting loss for me, but financially I didn't lose a penny. National flood insurance paid for the house and its contents," he explains, lounging on the beach. "And now that you're paying to replenish my beach [with taxpayer sand], it's possible that my neighbors and I will build our houses again and you'll insure us again. Thanks." He then flashes his trademark smirk for the camera.

While Stossel appears to be critical of the program (while still taking money from it), author Nicholas Sparks defended FEMA flood insurance in a September 19, 1999, *New York Times* editorial following Hurricane Floyd. His home, built on Bogue Banks, a North Carolina barrier island, has been repeatedly damaged by storms. Still, he argues that in a major storm surge the houses of poor people on the mainland would also be destroyed. "That's the real reason for the subsidy," he claims. "It protects large numbers of people who are not rich, whose homes are worth less than $250,000." Of course, those are the homes least likely to file repeat claims, being far less exposed to flood damage than properties on high value–high risk beaches like Bogue Banks, Westhampton, and Malibu.

A study put out by the National Wildlife Federation in 1998 confirmed that 40 percent of all federal flood insurance payments went to repetitive loss properties. For example, a commercial property on Fire Island, New York, collected $96,000 in 1991, $203,000 in 1992, $42,000 in 1993, and

$250,000 in 1994. A beachfront property on Topsail Island, North Carolina, collected $457,713 for losses from eight different storms between 1981 and 1996.

In the fall of 1998, James Lee Witt, the reform-oriented director of FEMA, gave a speech at the National Press Club in Washington, D.C., in which he proposed that insurance be denied to home owners who file two or more claims that total more than the value of their house and who also refuse to elevate or floodproof their homes or accept a buyout and relocation offer. The speech was immediately attacked by the National Homebuilders Association, and FEMA began a quick retreat from what could only be described as an already overly cautious proposal.

As early as 1982 Congress began to see the folly of promoting dangerous coastal development, so it passed the Coastal Barriers Resources Act (CBRA, pronounced "cobra"). The act excludes some 1.3 million acres of flood-prone undeveloped barrier islands and sand spits from federal subsidies. It does not mean property owners in these areas can't develop their land, only that they cannot bring in Uncle Sam as their real-estate partner.

"I travel the coast, and the CBRA units are not being developed and you have to ask why? Well, you can't get flood insurance and other federal money and that clearly suggests that federal insurance and beach replenishment and road and bridge construction must be encouraging development," says Woods Hole's Jim O'Connell.

Having passed some good legislation, Congress quickly went to work tampering with it. Beginning with the 104th Congress, the House and Senate passed dozens of "technical corrections" to CBRA in order to remove constituents' properties from CBRA's subsidy-free zones. These "corrections" relate to properties that were already developed or being developed in some way when they were included in the act. Typically, on the last two days before the 1999 Thanksgiving break, Congress rushed through a number of "technical corrections" restoring federal subsidies to properties in Delaware, North Carolina, and North Captiva Island, Florida, one of the wealthiest zip codes in America. The changes were requested by Republican Representatives Michael Castle of Delaware, Walter Jones of North Carolina, and Peter Goss of Florida.

About 12 acres on the exclusive Ocean Reef Club in North Key Largo and the 25-acre Pumpkin Key just offshore from it won earlier "corrections." The Berry family, which owns Pumpkin Key, wants to develop a dozen homesites on their small island property (nearby homes start at a million dollars each).

I call Terra Cotta Realty (Florida), which is actually located in Crystal Lake, Illinois. I speak with Bob Berry, Terra Cotta's vice president, and ask if I can visit Pumpkin Key. He says he'll call back and, when he does, tells me they would be happy to send me their congressional testimony, "but our attorney doesn't think it would be appropriate for you to visit."

I later run into Representative Pete Deutsch, the liberal Democrat from south Florida who got the two properties taken out of CBRA. "The fact is they shouldn't have been included," he tells me. "I went out there to Pumpkin Key and there are roads and plumbing hook-ups. And the Ocean Reef property was fully developed [and occupied by multimillion-dollar homes]. It was a technical issue. Based on the legislation it was a mistake." He looks around, seeming uncomfortable even in his nicely tailored suit. "Nine projects were looked at. I think the government opposed five exclusions. Ours were supported by the [Clinton] administration."

Deutsch turns away to talk to a reporter from the *Miami Herald* about Cuban exiles confronted by the Coast Guard in the water off Florida. He compares Fidel Castro to Serbia's Milosevic, America's main enemy at that moment, and goes on to complain about returning illegal immigrants to Cuba, "the last dictatorship in the Western Hemisphere." In south Florida speaking ill of Fidel Castro has never hurt a politician, nor has speaking out for real-estate interests.

I look at the House testimony of Bob Berry's father, George, who bought the low-lying island in 1973 and has since put $5 million into his private paradise (not counting $100 a day of cut flowers when his wife is there). More interesting, however, is the testimony of Terra Cotta's Chairman Tom Hayward in front of the Senate Committee on Environment and Public Works on September 22, 1998. What is interesting is that on the day he asked Florida Senator Bob Graham and others for passage of a law to restore the Berrys' development subsidies, Ocean Reef and the Keys were being evacuated because of Hurricane George. George left a billion dollars of damage in its wake, including $250 million to the Keys. Still, the Berrys had their subsidies reinstated.

"We still purchase federal flood insurance," Bob Berry tells me when we talk again. "We were looking into not purchasing it, but you can't get homeowner's, liability, or any other type of insurance unless you have it."

Tom Hayward joined the conversation. "CBRA would have impacted on our marketability. Without flood insurance you wouldn't be able to get a mortgage if you wanted to buy a house there [on Pumpkin Key]."

"We're not interested in being part of any book," Bill Hackelton, director of communications for Ocean Reef, tells me when I call to inquire about a visit. The 4000-acre gated community has 1600 residences, plus golf courses, tennis courts, hotel rooms, private restaurants, a small airport, and a large yacht basin. Incorporated as a private club in 1993, Ocean

Reef has membership rates of $117,000 and $300,000. The higher rate gets you discounts on food, beverages, and golf. Either one lets you buy properties ranging from one-bedroom condo efficiencies for $280,000 to $7 million and $8 million waterfront mansions. This has made it prime winter-golf habitat for Kathie Lee and Frank Gifford, Whitney Houston, George Bush Senior, various corporate CEOs, and the occasional Kennedy or two.

"Don't you sometimes hold conventions?" I ask.

"We hold conferences, not conventions. We don't invite retired postmasters in. We have senior management retreats and CEO meetings."

"So reporters aren't welcome?"

"Not unless you're a guest of one of our members."

Or go in with the service guy who scrapes the barnacles off their yachts. But like the Douglaston Country Club when I was a kid, once you've seen Ocean Reef it's hard to get too excited. It is a suburban dream writ large; wide meandering streets, apple-green golf courses, Spanish-accented rooflines (and service employees). The street traffic is mainly scooters and golf carts occupied by white guys wearing Lacoste shirts.

We swing past the marina and over to Snapper Point, where I walk to the water's edge and pull out my binoculars. Thirteen hundred feet away is Pumpkin Key. I could swim out to it, but the warm, soupy brown water of Card Sound doesn't look too inviting unless you're a manatee. Plus I'd be trespassing. But I can see there's infrastructure in place, a gray building sticking out of low-lying mangroves and a dock with a 70-foot yacht pulled up to it. My tax dollars at rest.

Despite the risks from hurricanes, rising seas, and golf carts blocking emergency vehicles, alternative models for coastal development do exist, even in Florida.

In the movie *The Truman Show,* Jim Carrey is raised in a utopian beach town that he discovers is really a domed TV set. Although it was altered for the film (computer graphics added a curved horizon, fake river, and tall buildings), the real-life town of Seaside, Florida, made a convincing set because it is something of a visionary beach town. Seaside is located on the emerald blue-green waters of the Florida Panhandle about halfway between Pensacola and Panama City.

When Hurricane Opal hit the panhandle on October 4, 1995, with high winds and a 15- to 20-foot storm surge, towns like Destin and Dune-Allen, just a few miles from Seaside, were devastated. But Seaside's 280 Old Florida–style frame homes came out unscathed. "They protected the dunes and the dunes protected them," explained Donna M. Dannels, the FEMA official sent to assess the damage.

Aside from building behind the beach's protective sand dunes, Seaside also used tough building standards designed to withstand 150-mile-per-hour winds, sank its foundation pilings deep into the ground, and planted lots of native trees and grasses to secure the dunes and buffer the houses. Cheap, modern building materials like vinyl and strandboard siding were rejected. "If it didn't exist before 1940, we didn't feel it was proven," says Robert Davis, the founder of Seaside.

The development he began in 1980 has grown to 400 beach cottages and shops on 80 setback acres of scrub, live oak, magnolia, and pampas grass, picket fences, and pastel-painted homes, many with steeply inclined tin roofs and screened-in porches, and all linked by rain porous cobble streets and sandy footpaths. Its "neo-traditional" style and low-impact design have won praise from *Time* magazine, *Architectural Digest*, and the Florida Audubon Society.

"As a kid I spent summers with my family in simple, expendable beach cottages near Panama City, where they later built bulkheads and put large buildings behind seawalls and you can guess what happened next," Davis says. "They've had to take some of those buildings down after the beaches went away. . . . It used to be everyone built well back from the beach. That was before federal flood insurance made stupidity feasible.

"What washed away [during Hurricane Opal] were slab houses built on bulldozed sand, really stupid construction. We'd all be better off to let hurricanes do what they do and take the proceeds from the insurance and buy somewhere else, but I suspect we'll just keep shoveling against the tide."

I ask him if he sees Seaside as part of a trend toward more careful coastal developments.

"This is less of an exception than it used to be, but calling it a trend may be overly optimistic. . . . I built Seaside not just for its environmental advantages, but because I thought it would be more pleasant to build low density. We've created land values two blocks from the beach that exceed the beach-front condos around here. People will pay huge amounts for access to unencroached-on wilds and beaches. One reason the more coastal protection the state of California enacts, the higher its coastal property values climb.

"Good developments are done with long-term greed as their operating principle," Robert Davis argues, "but there's still more short-term greed at work. We make piles of money and show our colleagues you can make money slowly and not rape and pillage the land. We try and share the idea of building lightly on the land [and by the water]."

The city of Hilo, Hawaii, is another example of how stepping back from the edge can work to the benefit of both local residents and the economy.

After losing dozens of people and its downtown waterfront district to tsunamis in 1946 and again in 1960, the city came up with a new plan called Kaiko'o ("rough seas"). Downtown commercial development was moved inland behind a 350-acre bayfront park and fish market. Today the landscaped park (designed to take the brunt of the next tsunami) includes Japanese gardens, picnic grounds, and a palm-tree Coconut Island. The creation of this recreational space also inspired a movement to clean up the bay and got surfers, boaters, and kayakers back into the water. Hilo, with its refurbished waterfront, became a popular destination for both out-of-state visitors and other Hawaiians who see it as the last traditional Aloha city. Real estate overlooking the park and bay now sells for far more than the old waterside properties.

While maintaining natural setbacks like these makes both environmental and economic good sense, there is a more troubling reason people choose to turn their backs to the sea. According to a recent report from the National Academy of Sciences, marine life in more than a third of U.S. coastal areas is being killed off by nutrient-fed algae blooms. At the same time there are more beach closures and coastal health advisories being posted than ever before.

"In 1987 and 1988 we had a disaster with our coastal economy because of beach closures, algae blooms, medical waste, and dolphins washing up dead on our beaches," recalls Representative Jim Saxton of New Jersey. "We made a lot of needed changes in response to that kind of point-source pollution. But a more difficult challenge is still ahead of us."

Chapter Eight

Flushing
the Coast

Surging rivers are carrying a stew of sewage, urban runoff, farm chemicals, silt, and debris. At least 10 swine-waste lagoons have flooded or burst.

—HURRICANE FLOYD COVERAGE, RALEIGH NEWS
AND OBSERVER, SEPTEMBER 1999

The beach is open. Only the water is closed!

—SURF-SHOP OWNER MICHAEL ALI, ON THE 1999 CLOSURE OF
HUNTINGTON BEACH, CALIFORNIA, DUE TO BACTERIAL POLLUTION

I'm diving through the warm blue-green waters of Conch Reef, 7 miles off Key Largo, Florida. What appears below looks like the underwater lair of some James Bond villain. Actually, it's a 48-foot cylindrical habitat called Aquarius, the last underwater research station in the world. It seems to hang weightless on its four steel legs, its yellow body rusting to orange in spots and encrusted with weedy growths being grazed on by roving schools of fish. On a platform just off the entryway is a white six-sided gazebo with an air pocket—Aquarius's temporary rescue hut should it need to be evacuated in a hurry. I swim by one of the habitat's round viewports. A scientist inside looks up from his laptop and waves, a can of salted peanuts by his side. I turn around and notice three big tarpon shadowing me. The largest fish has to be at least 150 pounds.

Running from the habitat across the sandy bottom and onto the reef is a spaghetti mat of cables, enough to power a Lollapalooza concert. In this case, the cables end at two acrylic domes and a tripod-mounted stereo camera rig. Under one of the domes is a brain coral, its respiration being measured as part of this month's mission to evaluate the condition of North America's greatest living reef.

Coral reefs are the most diverse ecosystems on earth (25,000 species and counting) and the nursery for more than half our tropical seafood including lobster, red snapper, grouper, and drum. They also act as coastal storm barriers and provide epic adventures for sport fishermen, kayakers, snorkelers, and, of course, divers.

Steve Miller, director of the University of North Carolina's National Undersea Research Center, which administers Aquarius, leads me under the skirt of the habitat. We slide through the low-slung, 50-foot-deep entry to the wet room, and pop up next to two researchers from the New England Aquarium about to make their way out onto the reef. A school of yellowtail snappers huddle discretely toward the back of the wet well. Waist deep in water, we strip our gear and climb up a short ladder with the assistance of Jay Styron, one of Aquarius's two live-aboard support staff (out of a crew of six). In his spare time, Jay, a former charter boat crewman, acts as the Clara Barton of barracudas, grabbing the ones that got away by stray line and leader and, holding them under one arm, removing the hooks from their mouths.

The inside of the habitat is cramped but comfortable with blue industrial carpeting, a lab, kitchen, and two sets of triple bunks. For Dr. Mineo Okamoto, a visiting scientist from JAMESTEK (Japan Marine Science and Technology Center), it's a dream come true. "In seven years studying coral from boats I could never do what I'm doing here," he says, his voice squeaky from the increased atmospheric squeeze on his larynx. Saturated with highly compressed air to compensate for the pressure, he and the other aquanauts will live inside Aquarius for eight days, scuba diving up to nine hours a day and spending much of the rest of their time monitoring their reef subjects, as they themselves are monitored from a wireless watch desk back at Mission Control, a gray stilt house in Key Largo. Mineo appears to be slightly giddy, not from nitrogen narcosis, that drunken feeling too much nitrogen in the bloodstream can give you, but from having stayed up all night reading the respiration rate of his coral as it appears on his laptop computer, which is secured by bungee cord to a narrow shelf on the far wall. He shows me a color graph running across its flat matrix screen that means nothing to me. He then hands me his business card. I apologize for not having thought to bring my own.

Unfortunately, as surface interlopers, Steve and I only have a little over an hour visiting time before the nitrogen beginning to build up in our body tissue might cause decompression sickness, or the bends. At the end of their stay the Aquarius crew will go through a controlled 17-hour readjustment to surface air pressure, with the habitat itself acting as their decompression chamber. After they are brought back up to ambient surface pressure, they'll step into a smaller lockout chamber, the space between the wet well and living area that includes a toilet, shower, and some camera shelving. Here they will be quickly repressurized to depth in order to return to the surface via the moonpool, as if they had just been down on a five-minute dive.

On a subsequent visit with a video crew from *National Geographic*, I check out the seascape surrounding Aquarius, and realize one of the reasons this undersea lab is located where it is. Despite the fish drawn to the habitat (as by any underwater structure, be it a shipwreck, oil rig, or barrel of nuclear waste), this reef is dying. Branching corals that once grew here remain only as skeletal sticks in bleached rubble fields. Many of the abundant rock corals are being eaten away by diseases that have spread in an epidemic wave throughout the Florida Keys. The names of the diseases tell the story: black band, white band, white plague, and aspergillus, a fungus normally found in terrestrial soil that can shred fan corals like moths shred Irish lace. The corals are also being smothered under sediment and algal growth linked to polluted runoff and are periodically bleaching white as a result of warming ocean temperatures.

"The story of this decade is one of coral decline," says Miller. Coral reefs, while grand in appearance, are actually fragile structures, living within a narrow range of clarity, salinity, and chemistry so that even a slight increase in the ocean's temperature or increased carbon dioxide input into the ocean from the burning of fossil fuels can cause stresses—like bleaching or more acidic waters that can slow the rate of growth of the corals (which strain calcium out of seawater in order to build their limestone castles). With some 60 percent of the world's coral reefs now losing productivity, it has become a global crisis and scientific mystery with one prime suspect. Reefs found to be at greatest risk of dying are those closest to human development. And no reef line in the world is closer to urban sprawl than the 126-mile-long Florida Keys reef.

With diving, boating, and recreational fishing offered at every mile marker, the Keys are a major contributor to south Florida's $15-billion-a-year tourism industry. Unfortunately, the Keys are also downstream from 6.5 million south Floridians and their urban and agricultural runoff, including runoff from Florida's huge (and hugely subsidized) sugar industry.

Nutrient-rich phosphorus and nitrate fertilizers, pesticides, and herbicides used in the sugar industry's 300,000-acre Everglades Agricultural Zone have been identified as major polluters of both the historically low-nutrient Everglades and the Keys' historically low-nutrient bay and reef system. Unfortunately, the politics of sugar have tended to play a dominant role in both Florida and Washington's response to the problem and even showed up in a little noted passage of Ken Starr's report on former President Clinton.

In section three of the Starr Report, titled "January–March 1996: Continued Sexual Encounters," sub-section "President's Day (February 19) Breakup," in which Bill tells Monica they can't go on, hugs her but doesn't kiss her, he is interrupted by a phone call. At this point the report interrupts its own rather salacious narrative flow to explain that "the President talked with Alfonso Fanjul of Palm Beach, Florida from 12:42 to 1:04 p.m. . . . The Fanjuls are prominent sugar growers in Florida."

So what did the President of the United States and the Cuban exile and owner of one of the state's top sugar companies, Flo-Sun, have to discuss for 22 minutes? Without Starr's power of subpoena (or Richard Nixon's taping system), we are unlikely to know exactly what was said, but it is indicative that Vice President Al Gore was to speak in Florida that afternoon, unveiling the administration's plan to restore the Everglades. The plan included a one-cent-per-pound cleanup assessment on sugar growers, a provision eventually dropped by the administration. A farm bill was also coming up for consideration in Congress, and while most federal subsidies to agricultural commodity producers were ended, sugar managed to hang tough (the White House did not support repeal of the sugar subsidy). That single subsidy brings Flo-Sun more than $64 million a year according to the General Accounting Office. Careful to cover all his bases, Fanjul worked as Florida's finance chairman for the Clinton-Gore reelection campaign in 1996 while his brother José acted as vice-chair of finance for Bob Dole's presidential campaign. The Fanjul family spread $774,500 in donations among both political parties in that election.

Still, the reef's problems cannot be traced to a single source of pollution such as sugar, which means there is no silver-bullet solution to save the reef. Along with agricultural sources of nitrate and phosphorous runoff, Florida is also home to phosphate mining companies and their waste waters. New construction and oily runoff from paved streets, driveways, and shopping malls throughout south Florida add additional silt, nitrogen, and hydrocarbons to the water, as does air pollution when it rains out. Studies of the Gulf of Mexico suggest that phosphorus and nitrogen runoff from the Mississippi and several Latin American rivers have also found their way onto the reef.

Nitrogen is essential for soil productivity, but too much of a good thing can be a bad thing, as anyone who has ever had a hangover will attest. According to various studies and recent reports in *Science* and *Scientific*

American, synthetic fertilizers developed after World War II and the burning of fossil fuels doubled the global nitrogen cycle between 1960 and 1990.

Along with natural nitrogen found in air, soil, and lightning, the added nitrogen input is too much for the land to handle, and the surplus is washed off into the world's rivers, estuaries, and oceans, where it ends up feeding giant algae blooms.

In addition to these global factors, more than 80,000 local residents and 2.5 million annual visitors to the Florida Keys rely on septic systems that flush untreated waste through the islands' porous limestone (the skeletal remains of previous generations of coral) and into nearshore waters, changing them from aquamarine to a lime Jell-O color. Key West, with the only municipal sewage plant on the islands, saw its beaches shut down during the summer of 1999 because of high bacterial counts as city workers dug up leaky pipes for replacement. At the same time, a four-cent-per-bed hotel tax goes not to environmental cleanup but to the Tourist Development Council, which spends $10 million a year encouraging more visitors to come flush the Keys.

Corals, in their most productive state, need clean, clear, low-nutrient waters to thrive. By contrast, algae loves sewage and other nutrients, and in their presence will bloom into a green, light-obscuring soup that sucks the oxygen out of the water as it decays, killing off massive numbers of reef fish, while suffocating the living coral polyps.

"You can emphasize but not get bogged down by the complexity of the reef environment," suggests Steve Miller. "What we have to do is consider those things we can control that may have a negative impact on the reef. We can control overfishing [of lobster, parrot fish, and other grazers that target algae], we can control nearshore pollution, and maybe some of those nutrients. But it can't all be done locally. What Midwest America dumps on its streets and adds to its soil ends up in the Mississippi and the Gulf of Mexico, which ends up on our reefs. People have to begin to understand that you may live in Iowa but you're also living on an ocean planet."

He says this with a kind of affable gentility. Steve, with his bearded smile, twinkling chestnut eyes, and slightly nerdy oblong glasses, is, after all, living the good life. He manages the National Underseas Research Center from a Key Largo canal complex that has the feel of a water rat frat house, but with the added adult pleasures of home, wife, and family just down the road.

He shows me the extra decompression chamber parked in the flood zone under the gray stucco stilt house. I look in through the window at its two molded fiberglass benches that seat seven.

"Looks pretty uncomfortable."

"When the alternative is death, it's comfortable," he says, and then grins. "Hey, that's pretty good. You going to use that in your book?"

Late that evening I'm kicking back at the watch desk upstairs with several of Steve's staff, including Craig Cooper, operations director for the program. A big man with blond hair blending to a white beard and cool blue eyes in a sun-reddened face, Craig gives an impression of solid competence, which is what you want when facing the frontier challenges this former oil rig diver has to face every day. Right now he's video conferencing with Jay inside the habitat, discussing a pinging sound on their shared computer net and a carbon monoxide default on one of the new sensors they installed. They agree that the sensor needs to be reset and watched.

"We have to figure these things out as we go along," Craig explains. "There's no habitat store to go to and pick up an extra whatever." Two nights earlier Aquarius's air conditioner had failed, and, instead of pulling the huge unit, Craig improvised a way to get glycol (antifreeze) down to the habitat in a sealed hose.

"Last year the power went out and I had to decompress inside the habitat for 17 hours with this scientist. The temperature was 95 degrees with over 95 percent humidity. That was the worst decompression I ever went through."

Bottom time in the habitat can also get boring for the regular crew, so they've found ways to amuse themselves. When recreational divers used to come into the area Craig would play the theme music from the movie *Jaws* on the underwater hailer.

"Some of them were alright and came to the window and would pretend to get pulled under. Others were clearly nervous when they heard it."

I ask him how he got into the business. He says he read a magazine ad for commercial diving and ended up in Houston for six months at a commercial dive school, training for the most dangerous profession in America.

While working for Taylor Divers in 1977, he was offered $1000 a day to work 1000 feet down on the construction of Shell Oil's Cognac platform but turned down the offer. One diver who volunteered for the job told him he saw 1200- to 1500-pound groupers down there, "real sea monsters."

For $500 a day he did saturation diving in 600 feet of water, rewelding pipeline housings with bad swivel joints. "We'd pull the pipes off the bottom and drop these huge hydraulic frames and put a 16-by-16 housing over them and dig our way in underneath, working out of a diving bell. Once inside the hut you'd take your [tethered] dive helmets off to work, but you worried because the ship up top was using dynamic positioning to stay in place and if that fails, your helmet will be dragged off, leaving you behind."

He says he had a roommate and several friends who were killed while diving, including a buddy in the water with him at the fiery 1979 Ixtoc oil platform blowout in the Gulf of Mexico. Those kinds of experiences have made him a stickler for safety and quick response. His go-fast boat tied to the quay outside can be at the habitat in 20 minutes. I ask if there have been any close calls with Aquarius.

"Four of us got stuck on the top barge [that was later replaced with a buoy] during that '94 storm named Gordon," he recalls. "We were stuck on the barge for 28 hours, then had to evacuate the habitat in 15-foot seas. The Coast Guard wouldn't help cause they said it wasn't a life-threatening situation, but we needed to decompress the scientists and get them out. Our generators had gone down and one had caught fire. So, they were down there using battery power for the CO_2 scrubbers, with no functioning air conditioning. We finally got back out to them on a 50-foot catamaran owned by a local dive company. We established a down line in zero visibility by crawling down the umbilical [power and communications line] and then taking the rescue line back up to the cat. We'd send two men up the line and then motor toward them and shut down the engine. As soon as you powered down you'd start turning broadside to the waves, though. A few times I thought the boat was going to go over. Six of us would grab the divers out of the water and power back up before the wave took us. We used everything but gaff hooks hauling those guys out of the water like tuna.

"Hurricane George also had 5-knot currents 80 feet down. It broke one of the four support legs on the habitat. The buoy was recording 28-foot waves, and the top of the habitat is only 34 feet down, so it was lucky no one was aboard that time. We stayed here on the canal rather than evacuate the island. The water was two and a half feet above the dock with 90- to 100-mile winds. That was another close one."

It was not the wildness of life on the Blue Frontier but the absence of it that Dr. Nancy Rabalais of the Louisiana Universities Marine Consortium first noticed. Diving on an abandoned oil rig 25 miles off the coast of Louisiana, she was surrounded by shoals of fish, including sheephead, drum, and yellowtail snapper. Going deeper, she passed through a kind of hazy shimmering layer about 30 feet down and then, looking around, noticed a strange change. Suddenly there was nothing but empty murk as far as she could see.

Sometimes while diving you can sense a presence, something lurking in the water just beyond your field of vision. It is a not uncommon feeling, whether you are looking out into the borderless azure blue of the open sea or swimming cautiously through brown coastal waters that can reduce visibility to feet, sometimes even to inches.

Fish will often scatter moments before a big shark or other silent predator swims into view, sending your heart racing. Nancy Rabalais looked up but the fish above were still visible through the murk, schooling and seemingly unconcerned. She dove on looking for life in the normally teeming water but found none. On the muddy bottom 70 feet down, however, she did find a lifeless crab and then a brittlestar and white seaworms that had crawled out of their holes in the mud before they stopped crawling at all. That was when she knew, with a certainty that sent a shiver racing across her flesh underneath her wetsuit, that her calculations and studies were all correct—she had finally entered the Dead Zone. A decade later much more is known about the nutrient-fed Dead Zone in the Gulf of Mexico, thanks to the work of Dr. Rabalais and her colleagues.

The Dead Zone, which is more than 7700 square miles or about the size of New Jersey, lurks unseen below the gulf's surface, is seasonal in appearance, and amoebae-like in structure stretching and writhing like some stealthy creature, every spring and summer. First studied in the 1970s, it doubled in size following the great Mississippi flood of 1993. In 1998 it shrank in length but expanded in depth. In 1999 it was both deeper and wider than ever before, but in 2000 it shrank again.

What makes a dead zone is lack of dissolved oxygen (DO) in the water. At least 5 milligrams of DO per liter makes for good life-supporting water. When it drops to 2 milligrams per liter the water is called hypoxic and is a risk to shrimp, fish, and shellfish. "Anything that can't move out eventually dies," according to Rabalais. Below 1 milligram the water is anoxic or without any means of supporting life. The whole deadly process of creating this nutrient-rich, oxygen-poor water is called eutrophication. As the scientific understanding of the gulf's eutrophication has become more focused, the economic consequences have vastly expanded.

The Mississippi watershed drains 41 percent of the continental United States into the Gulf, including 52 percent of farms that during the 1950s and 1960s became increasingly dependent on synthetic chemicals. These commercial fertilizers and pesticides, promoted by the government, American Farm Bureau Federation, and Agricultural Chemicals Association (since renamed the American Crop Protection Association) boosted production but undermined the long-term viability of the soil.

Today, in states like Illinois and Iowa, corn farmers, caught up in a cycle of chemical dependence, apply 150 pounds of fertilizer per acre. Corporate feedlots for cattle, hogs, and chickens also generate tremendous amounts of largely unregulated methane, ammonia, and nitrogen. Every spring, rain and snowmelt deliver much of this nutrient-rich brew (along with urban and industrial runoff) into the Mississippi River. The river discharges its spring flow of freshwater across the top of the saltier gulf, where, warmed by sunlight and fed by surplus nitrogen and phosphorus, massive algae blooms appear. They attract tiny crustaceans called copepods and other needlepoint-sized critters that graze on the algae. The dead algae, along with copepod waste, sinks to the bottom to become food for bacteria. In the process of feeding, these bacteria suck up most of the oxygen, dropping the water's DO levels to around 2 milligrams. Then, in a final voracious feeding frenzy, like late 1990s NASDAQ investors buying Internet stocks, they consume the last oxygen, suffocating even themselves.

To date, the Dead Zone has had mixed impacts on the gulf's sea life and billion-dollar seafood industry that represents about 40 percent of U.S. production. Benthic communities of seaworms and shellfish (that provide prey for other creatures) are able to recolonize after the seasonal environmental insults subside. Commercial shrimpers have been able to target the edges of the Dead Zone, sometimes benefiting when great swarms of shrimp flee directly into their nets. But many worry that if the Dead Zone continues to grow, there will be no place left to feed or to flee.

There have always been some natural algae blooms and small hypoxic areas resulting from siltation and plant runoff in the Gulf of Mexico, whose rich brown waters have long been known as the "fertile crescent." Mobile Bay, Alabama, for example, flushes some 50 billion gallons of water a day into the gulf, making it the fourth largest freshwater flow in the nation (after the Mississippi, Yukon, and Columbia Rivers). Records going back to the 1860s document periodic hypoxia events, called "jubilees" by local residents.

"That's when the shrimp and bottom fish are driven out of the water and throw themselves onto the land to get some oxygen, providing a free feast for the locals," explains Jonathan Pennock of the University of Alabama's Dauphin Island Sea Lab. "People put bells on the beaches that they can ring when this happens."

Pennock, who has silver hair, a trim mustache, and somewhat fastidious manner, helped write a 1999 University of Alabama (UA) report funded by the Fertilizer Institute that focused on contributing factors to the gulf's Dead Zone other than fertilizers. Among the factors highlighted were dredging of navigation channels, loss of wetlands, and unusually heavy rains putting more carbon into the water.

While not generally well received by other scientists studying the problem, the report was highlighted in a *Forbes* magazine article titled "Hypoxia Hysteria," penned by Michael Fumento, a conservative think-tank author. *Dakota Farmer* magazine also found the report highly persuasive, as did the Farm Bureau (a not-for-profit advocacy group and insurance company heavily invested in agrochemicals).

"Nancy [Rabalais] and me are being put in opposite positions by the media, which is unfair and putting our friendship at risk," worries Pennock. "People with the Farm Bureau and some other farm interests misused our information, claiming the UA report said carbon was the issue. But in my mind there should be movement to reduce nitrogen inputs upriver but also keep looking at these other inputs. . . . I think in 15 to 20 years from now farmers will do all they can [to reduce nutrients], and the major inputs will be from coastal growth, from you and me using autos, fertilizing our lawns, spraying rose bushes. We're still a notch away from that but it's going to happen."

To get to Cocodrie, Louisiana, you take Highway 56 out of Houma past the airport, oil tank farms, and sugarcane fields, the shrimp and off-shore work boats tied up by the canal, past Dan and Cyndi's Minnows and Our Lady of Prompt Succor. Pretty soon you're driving past houses and house trailers jacked 15 feet off the ground, with wetlands and water all around and (if you like good music) the Ragin' Cajun playing on your radio.

Cocodrie is fractured French for "crocodile." It may not be the end of the world but when you think you're near, look off to your right and you'll see what appears to be a small tan airport on concrete pylons. This is the Louisiana Universities Marine Consortium, or LUMCON. Off to one side, rather than commuter jets, several research vessels are docked, including the 105-foot steel-hulled RV *Pelican*.

"I have two rules for the media—film me before I get wet and no butt shots," says Nancy Rabalais. But when I later look at newspaper and TV clips featuring Dr. Rabalais, I'm reminded how you just can't tell the media what to do.

At 49, Rabalais is slim with wavy brown hair, bright green-brown eyes, and a sweet tentative smile behind a Texas twang unaffected by years in Louisiana. I mention Jon Pennock's statement that his work has been misused by farm interests.

"Jon was pretty naive to take their money and not think they'd use that report how they wanted," she replies.

Nancy Rabalais first began studying the gulf's nutrient runoff problem back in 1985 along with colleagues Gene Turner (whom she later married) and Bill Wiseman. She pulls out offshore charts, showing me research transect sets (study tracks) going back 15 years. Most of the lines are compass straight except for the first two, which curve oddly.

"That's where we were trying to stay close to the oil platforms," she grins. "The first year out I had a 21-foot Boston whaler to work off of, which I kept close to the platforms, so if the weather got rough we'd have someplace to run to. Then we got a 45-foot boat, and later began doing our summer research cruises off the *Pelican.*"

Like many marine scientists, she creates her transects by taking water samples at a series of established sites lowering a CTD (conductivity, temperature, and depth) rosette on a winch. The rosette is a long, cagelike array of pipes and fittings holding CTD instruments. She also measures chlorophyll, suspended sediments, and active (light) radiation, and pulls bottom samples. In addition to a permanently moored instrumentation package, she oversees monthly day cruises and an extended summer research cruise that includes boat-based and dive operations.

"In December 1995 we gave science presentations on our findings in New Orleans, and there was a large agricultural presence there, but also scientists from the U.S. Geological Survey and fisheries people reporting the same findings we were reporting, so we were validated at this meeting."

The next year her team was asked to give presentations in Davenport, Iowa. "There were more ag [agricultural] interests there; the Phosphorus and Potash Association, the Fertilizer Institute, the Farm Bureau. . . . I very seldom turn down requests to talk, so I've become a kind of target for them. I've been told I'm too much of an advocate. But I don't state policy. I say what's happening in the water. I'm talking to the Farm Foundation soon in Corpus Christi. I'll talk to anyone who's interested because this is important."

"Excess nitrogen in our coastal waters starts a dangerous chain of ecological events that is reducing biodiversity, destroying sea grass, and contributing to a host of other environmental problems," agrees Robert Howarth, a Cornell University professor who led a National Research Council team that in April 2000 issued a report calling for a coordinated national plan to reduce nitrogen and phosphorus pollution of U.S. waters.

I ask Rabalais if criticism of her intensified following a White House proposal that Midwest farmers reduce their use of nitrogen fertilizer by 20 percent and restore five million acres of wetlands to absorb excess runoff.

"I think a lot of farmers are really concerned about their land and the impact they have on it. The stuff you hear, the attacks, are mostly out of the lobbyists and the big agricultural groups. We were on a radio show in Illinois and had a farmer call on his cell phone from his fields and talk about the problems he faces in terms of the economic pressures and costs for chemical treatment, and he was really caught between a rock and a hard place and we didn't have any answers for him."

But Tony Thompson, a fifth-generation Minnesota farmer with 1850 acres of corn and soybeans, has at least a few answers.

"We farmers are often applying more nitrogen than we need. We can reduce nitrogen without suffering yield loss," he told *Newsweek* magazine in October 1999. Thompson, like many innovative farmers from Maine to California, is combining traditional ecological methods of soil management, including planting buffer strips along rivers, protecting some prairie and wetlands, and raising organic crops, with high-tech methods like Global Positioning System (GPS) satellite mapping, which targets where and how to use herbicides and fertilizers sparingly. The result is far less waste in the form of nutrient runoff from Thompson's fields.

Unfortunately, this kind of careful, targeted approach tends not to work well for large-scale industrial farming including the factory farming of hogs and chickens that exploded in the 1980s and 1990s under the sponsorship of Murphy Family Farms, Tyson, Perdue, and other large agrocorporations.

The worst-case scenario of agroindustrial pollution may be taking place in North Carolina. It was there, in 1990–1991, in the warm, slow-moving river waters of the eastern coastal flood plain, that Dr. JoAnne Burkholder first discovered *Pfiesteria piscicida,* the so-called "cell from hell" aquatic microbe. Pfiesteria, which spends most of its existence buried in the mud as a harmless cyst, has a chameleonlike life cycle of at least 24 stages, in one stage disguising itself as a green plant and during four stages turning into a deadly predator.

Stimulated by excess nutrients, pfiesteria in its most toxic stage can strip the flesh off fish within 3 minutes, leaving fatal lesions the size of half dollars. Humans who've been exposed have lost consciousness and suffered memory loss, fatigue, nausea, vomiting, diarrhea, bloody sores, painful muscle contractions, and respiratory distress, among other symptoms. While working in a poorly ventilated lab, both Dr. Burkholder and her assistant were exposed to airborne pfiesteria spores that led to months of debilitating neurological disorders.

Dr. Burkholder's work in the early 1990s paralleled the growth of the North Carolina hog industry, which grew from about 2 million to some 10 million swine between 1990 and 1997. With hogs producing four times the amount of excrement as humans, this meant upward of 20 million tons of annual hog waste that the factory farmers keep in huge untreated lagoons or spray on fields incapable of absorbing all the manure they are hit with. Along with runoff, additional waste vaporizes as ammonia and then rains out into the water.

In 1995 a hog lagoon spilled 22 million gallons of waste into the New River. Millions of fish died there and in the nearby Neuse River, which was also struck by hypoxia. Along the coast, massive pfiesteria outbreaks killed as many as a billion small and medium-sized fish. At the same time,

Burkholder's credibility came under attack from the governor's office and the state Department of Environment, Health, and Natural Resources. These often personal attacks took place even as her research was being published in *Nature* and other prestigious science publications.

The reason for the attacks became apparent when the *Raleigh News and Observer* ran an investigative series by reporters Pat Stith and Joby Warrick. The series, "Boss Hog," revealed how the state's powerful hog industry (including former state senator and hog operator Wendell Murphy) convinced the state government to go easy on environmental controls for the sake of corporate profits. The newspaper series went on to win numerous awards, including a 1996 Pulitzer Prize.

In 1997 Burkholder's reputation was further redeemed when she became the subject of an admiring book, *And the Waters Turned to Blood* by Robert Barker, and was asked to advise the State of Maryland on how to respond to a new pfiesteria outbreak. That summer pfiesteria-linked fish kills were reported on several Maryland rivers that feed into Chesapeake Bay. While the "cell from hell" never entered the bay proper, consumer fears resulted in some $40 million of lost seafood sales for Chesapeake Bay fishermen.

Maryland's governor Parris Glendening responded to the crisis by committing his administration to reducing nutrient runoff into the bay and expanding regional and national efforts to restore its water quality. In 1999 the states of Maryland, Virginia, and Pennsylvania and the District of Columbia agreed to a 10-year plan to reduce all "runoff of harmful nutrients and sediments" in the Chesapeake's watershed—although how they will do this remains unclear.

One of the most documented sources of nitrogen and phosphorous runoff into the bay is the huge chicken production complexes on the Delmarva Peninsula (shared by Maryland and Delaware). There contract growers get their feed and chicks (some 600 million a year) from major poultry companies including Tyson, Perdue, Allen Family Foods, and Mountire Farms. When the chicks have grown to full-sized broilers, the companies pick up the birds, leaving the contract farmers with 750,000 tons of manure to dispose of.

The EPA has called for "comprehensive nutrient management plans" for these large-scale "animal feeding operations." The idea is to treat factory farms the same as municipal sewage plants and other point sources of pollution. As of 2000 the EPA plans remained voluntary and were not expected to go into effect before 2009.

In Maryland and elsewhere around Chesapeake Bay state legislators called for more immediate nutrient reduction that would also make the major poultry companies legally responsible for disposing of, excuse the expression, all their chicken shit.

"What about all the septic tanks and people fertilizing their lawns and municipal sewage plants? How come nobody's looking at them?" complains James Perdue, president of the nation's second largest poultry company. Which brings up a related question. What to do about the growing number of Americans embracing victimhood and refusing to accept personal responsibility for their actions, instead of thanking God for the opportunities this country affords them?

In a three-part series, "Poultry's Price: The Cost to the Bay," *Washington Post* reporter Peter Goodman documented how the poultry industry has responded to the threat of legal liability by hiring top-dollar lobbyists, including the former state attorney general of Delaware and former aides to the Maryland senate president and to Virginia Senator Chuck Robb. These providers of desired services for money not only managed to kill any and all legislation that would affect the companies financially but also got Delaware to reimburse the chicken moguls for half the cost of a phosphate-reducing feed additive that Maryland legally requires them to use.

Meanwhile, back in North Carolina, the state finally began inspecting its factory farms, finding hundreds of health and safety violations, poisoned drinking wells, and dead waterways. In the spring of 1999, reversing his previous course of inaction, the Democratic governor Jim Hunt proposed phasing out the state's hog waste lagoons and removing them from the coastal floodplain. Unfortunately, his proposal came a storm season too late. Within months Hurricane Floyd struck.

The flood that resulted sent a witch's brew of animal and human waste, dead livestock, and toxic chemicals into the 1700-square-mile Pamlico Sound, the country's second largest estuary after Chesapeake. Algae growth accelerated and decayed at 10 times the normal rate, creating a massive new dead zone more than 350 square miles in size with anoxic DO levels of 1 part per liter.

Dying coral, oxygen-starved dead zones, and a microscopic cell from hell are only three of a growing list of science fiction–like horrors from the flushing of our coasts. Back in 1995 Woods Hole put out a federally funded report on the growth of harmful algal blooms (HABs). These include naturally occurring red and brown tides containing toxic algae that have, quite unnaturally, doubled in number and range since 1972. HABs now threaten every coastal state and territory on the Blue Frontier.

In 1995 more than 300 endangered manatees were wiped out by a red tide of one-celled dinoflagellates off the coast of Florida. In 1996 hundreds of dolphins and brown pelicans washed up dead along the Mexico–California coastline, killed by toxic one-celled diatoms. In 1997 South Padre Island, Texas, was hit by a red tide that killed millions of fish and sent tourists fleeing with hacking coughs from poisonous algae aerosolated by the surf. In 1998 more than 400 sea lions died in California's Monterey Bay after eating anchovies that had consumed a toxic algae. In Florida doctors diagnosed 13 people with symptoms associated with exposure to pfiesteria. The Centers for Disease Control labeled their condition estuarine associated syndrome (EAS), because all the patients lived or worked around coastal estuaries. In 1999 red tides again hit Florida, causing new fish kills and eye and respiratory irritations for beachgoers.

In Alabama in 2000 there was a meeting of scientists to discuss possible links between nutrient runoff, global warming, and global increases in jellyfish blooms. The biggest jelly bloom increases in the United States were found in Chesapeake Bay, the Gulf of Mexico, and the Gulf of Alaska. The big problem for Alabama was a dramatic increase in the number of stinging jellyfish just off the beach at Gulf Shores and other tourist destinations. Ouch.

Monty Graham and other scientists at the Dauphin Island Sea Lab believe that increased nutrient loading in the gulf is feeding zooplankton that in turn feed nonstinging tunafore jellies that now provide plentiful prey for their larger, more dangerous cousins, the stinging jellies.

In the Pacific Northwest more frequent appearances of toxic dinoflagellates have increased cases of diuretic shellfish poisoning (DSP) and potentially lethal paralytic shellfish poisoning (PSP).

A Seattle-based fisheries cop provided me a vivid demonstration of what PSP looks like. He grabbed his throat so that his face reddened, filled his mouth with his tongue as if it were swollen, and began choking, gurgling and flopping around on his desk quite convincingly. It made me much more careful about when and where I buy my oysters.

One place I will not buy shellfish is on the beach in Los Angeles. Santa Monica Bay's waters have improved, thanks to sewage treatment upgrades completed in the 1990s (that were originally ordered in 1972) but are still not what you would call swimmer friendly. For years Los Angeles argued that because it could pipe its wastes directly into deep water off its sloping coast, it deserved a federal waiver from the Clean Water Act, which requires cities to decontaminate and filter human waste water. Eventually, Los Angeles agreed to upgrade its four treatment plants after high bacterial counts and human viruses were discovered in coastal waters off southern California and at some of the area's most beautiful beaches including Rincon, Topanga, Malibu, and Laguna.

In 1999 Huntington Beach, just south of Los Angeles was closed to swimmers and surfers for most of the summer as a result of high bacterial counts. "The Beach is open. Only the water is closed!" argued surf-shop owner Michael Ali, one of hundreds of retailers who saw their beach businesses hammered.

Immortalized as "Surf City, USA" by the singing duo Jan and Dean, Huntington Beach has had more than its share of environmental insults. In 1990 it was hit by a major oil spill inspiring local surfer activists to protest, "No way, Dude! We don't want your crude!"

In 1995 its Republican surfer Representative Dana Rohrabacher played a key role in an attempt to defang the Clean Water Act. House Resolution 961, labeled the "Dirty Water Act" by its detractors, would have, among other things, eliminated EPA's water-quality monitoring program and exempted sewage discharges into the ocean from existing clean-water standards.

In September 1999, just as Huntington Beach's bacteria counts declined, more than a thousand used hypodermic needles washed up on its shore, restricting a planned annual beach cleanup for children and adults. Not surprisingly, visits to beaches in the Los Angeles area have declined by 50 percent since 1983, reflecting public unease with the area's water quality. Despite sewage cleanup, people are justified in their concern.

A 1996 study of Santa Monica Bay found that one out of 25 people swimming near storm-water drains became sick with pollution-related illnesses, mainly stomach flu and upper respiratory infections. Their risk for infection was at least 57 percent greater than for people who avoided those sections of the beach.

One positive effect of the Santa Monica Bay study was passage of a new California law that requires regular water testing and public warnings at all of the state's swimming beaches—one of the toughest such laws in the nation. Still, the waters flowing from the concrete-jacketed Los Angeles River and Los Angeles's 1265 miles of storm drains continue to pose an ongoing threat to human health.

"Storm water is lighter than seawater and creates a plume, a kind of liquid lens that sits on top of the saltwater and doesn't mix well. Most aquatic life is not in the top of the water. People swimming and recreating are in the top of the water, which is why this is such a human health threat," explains Judy Wilson, the straight-talking director of the Los Angeles Bureau of Sanitation.

"It's much easier dealing with nonpoint pollution if you know it's from a dairy farm because you can deal with the cow manure, but when you're dealing with a city of nine million people, all with extremely bad habits, it's a lot more difficult," she says.

"It's important to understand why we have the worst urban runoff in the country. It has a lot to do with our geography and climate. We have the Santa Monica Mountains to our north. In the winter, rain runs down to the coastal plain where it used to flood all the time. To address the flooding problem the Army Corps used concrete and straightened out the lines of the L.A. River to get this water to the ocean as rapidly as possible. We now have an underground freeway taking this water to the sea. Our beaches basically are the dumping ground for our storm drains. We call the first big storm of the winter the first flush. That's when all the paper, plastic, and everything that was collected all year in the storm drains just gets whooshed down to the ocean, and you get tons of trash on the beach, along with oil and grease that's collected on the freeways during the dry season, and also your dog poop, the chemicals used on your lawn, everything people use to wash their cars runs down the driveway and finds its way to the ocean.

"It's very difficult to educate people about this issue," she admits, "especially in such a diverse community with people speaking something like 43 languages.

"Remember when we started talking about seat belts, how they reduce deaths and we also talked about tobacco not being good for you? It took a whole generation to realize that if you don't use seat belts you'll fly out of your car and if you chainsmoke Camel unfiltered you'll probably die of lung cancer. So our first need is to make sure the public knows what's going on.

"We're also trying some new technologies. Catch-basin inserts [that trap waste material] cost about $1000 apiece, but when you have 35,000 catch basins, you're talking about a $35 million program. The very scope of L.A. means anything you do is expensive. But we don't have the money. We receive $28 million a year from a storm water pollution abatement [property tax] fund and half of that goes to flood control. One expert quoted in the *L.A. Times* said we should be spending about $500 per capita on storm water cleanup. We spend $5 per capita. If we spent $500, that would be a $1.5-billion-a-year program and then maybe we could get our arms around this problem. Right now we're running four sewage treatment plants including Hyperion, one of the biggest in the country, for $660 million a year. So, clearly the urban runoff problem is not something that local government can handle on its own. We need state and federal support. We can't do this on $28 million a year. That's just a joke."

There are a few things Judy Wilson has been able to handle on her own. In 2000 she got the Los Angeles City Council to approve a project in which storm drains known to have high bacterial counts will have their water flows diverted from the beaches to the city's waste-water treatment plants.

"The thing is we just have to keep all the bad news on the front page," she smiles without humor. "You [reporters] have to keep writing all these horror stories, I'm serious. That's the only way we're going to be able to move our programs forward.

"Our waste water plants are no longer the problem," she argues, "but I can't understand how to get at all the rest of it. Let's use copper as an example. Copper in our effluent [sewage] pipes has come down dramatically but copper in the water has climbed astronomically. That increase is all from urban runoff. We need to go back to the automobile industry and say to them, listen, copper brake linings [on L.A. freeways] are resulting in pollution. You need different ways of doing brake linings, different types of products that don't break down in the environment. But that's a hard group to lobby, so people are either going to have to get awfully sick, or there's going to have to be a huge new green [or blue] power movement in this country, or we'll have to find huge additional new revenues. Education and best management practices won't do it. It's going to take action against manufacturers, federal law and regulation. But first we need to get people to recognize the problem, we need good science to understand what's going on in the water, and we need public financial support."

If history provides lessons, the Clean Water Act could serve as a model for how to mobilize that support. After its passage in 1972 many politicians realized that the new law was not just about plugging the industrial pipelines polluting America's waterways. It was also a chance to launch new public works projects for any district that had an outmoded municipal sewage plant or outfall in need of upgrading. It provided politicians a chance to cut ribbons and be seen as friends of public health and the environment. As a result, Congress has been willing to allocate more than $40 billion in grants for new sewage systems over the past three decades. More work still needs to be done to get rid of outmoded combined storm and sewer systems such as exist in Washington, D.C., which back up and pollute local waters every time there is a heavy rain. Ponce and other towns in Puerto Rico have also been left behind because of EPA failure to enforce the law. Still, dramatic overall improvements have taken place. Water quality has been restored and wildlife and recreational opportunities expanded for thousands of the nation's lakes, rivers, and coastal communities.

If many of these gains are now being reversed by the poisonous nutrient-heavy flushing of our coasts, we at least know that realistic models of restoration are possible. All that is needed is the political will to fight for our watersheds and our Blue Frontier.

Senator John Kerry of Massachusetts is one of a handful of Washington politicians who understand the frontier values of America's living seas. An ex-navy combat man, antiwar veteran, and dedicated windsurfer, the tall silvering politician has been jokingly referred to as "a new wave Democrat." He has helped get Boston a $2 billion waste-water treatment plant on Deer Island in Boston Harbor. By improving the city's water quality, the Deer Island plant has also spurred a multibillion-dollar physical and cultural revival of the historic harbor and its adjacent coastline.

I catch up with Kerry at a Trade Center luncheon overlooking the harbor on a bitter cold winter day in January, with drifts of accumulated snow blowing between the slate-colored water and an achingly bright cobalt blue sky.

He is here to encourage some 300 businesspeople to give their time and money to the restoration of the state's wetlands. He grabs people's shoulders, gestures expressively, and tells a slightly shaggy story about General Francis Marion, the "Swamp Fox" who fought the British by retreating into Carolina's salt marshes to conduct guerrilla warfare, using the calls of swamp animals to signal his friends and confound his enemies. "And that," Kerry concludes with an engaging grin, "is how we owe our freedom to the wetlands." The crowd chuckles and applauds appreciatively.

"The Clean Water Act and past cleanups have lulled some people to sleep," he tells me after the speech. "Now we have to rekindle a sense of emergency. When you look at what's happening downwater in our bays and estuaries and salt marshes and how at risk they are from contamination, we need to re-educate people and re-energize people."

Kerry has submitted a bill to Congress called the Coastal Stewardship Act, which would direct 10 percent of offshore oil drilling royalties, at least $250 million a year, to a newly created Ocean and Coast Conservation Fund. Forty-two million dollars of that money would be targeted at controlling nonpoint pollution, a modest start. A larger conservation and reinvestment act proposal would provide at least a billion dollars a year for coastal conservation, some of which would be used for controlling polluted runoff.

"Leadership can put this on the table, and there's a huge constituency waiting to act on it," Kerry believes. "I grew up on the ocean and spend as much time as I can out on Buzzards Bay [Cape Cod]. If I didn't love it as well as I do, I probably wouldn't be as involved as I am. It's my touchstone, my sense of being and place to re-energize. But it's also a fragile system that needs to be capable of regeneration for itself."

As Kerry and almost every maritime student and fisherman in America knows, the continued flushing and fertilizing of our coasts could easily spell the death knell for our oceans' living marine resources.

Scientists like Steve Miller, Nancy Rabalais, and Joanne Burkholder have sounded the alarm as clearly as any patriot scout—"One if by land, and two, if by sea." It is now up to America to heed their call.

Of course, when it comes to the Blue Frontier's once vast and diverse fisheries and the birds, bears, marine mammals, and people who depend on them, runoff is not the only threat. For lobsters, swordfish, shark, salmon, and tens of thousands of other species, coastal pollution and habitat destruction, as bad as they are, are only half of an increasingly deadly equation.

Chapter Nine

The Last Fish?

My concern's for the fishermen facing those high seas. I haven't seen a Steller sea lion that's voted for me yet.

—Representative Don Young, May 20, 1999

Too bad fish can't vote. But then neither can our children or our grandchildren.

—Dr. Sylvia Earle, June 12, 1998

We meet Scotty Doyle by the Staten Island Ferry Terminal at 3:45 on a cold, dark drizzling morning. He introduces us to Tommy Graham, a short, swarthy New Yorker. Scotty is taller with curly gray hair under a bill cap, thin frame glasses, and green eyes. He also wears the bulky shape of a Kevlar vest under his shirt and black shell jacket and carries a 10-millimeter automatic, marking him as a fed. He's a National Marine Fisheries Service (NMFS) enforcement agent, but don't call him a fish cop. He doesn't like that. Tommy, in his worn khaki uniform, is with New York's Department of Environmental Conservation (DEC).

We approach the Fulton Fish Market across misty, rain-slick streets. The fish market used to be heavily mobbed up until Rudolph Giuliani, New York's mayor at the time, set up some controls. There's a mobile white inspection booth set on a corner where dozens of tractor-trailer rigs, box-refer trucks, and vans are parked waiting below the FDR Drive, within stone's throw of the Brooklyn Bridge and the East River. Company names are written on their sides: Tico Transport, Ameri-Cana, F&B Mussels, Ecuadorian Line.

Trucks coming from different locations used to have to pay kickbacks to different mob families in order to get their product offloaded. With city inspectors checking manifests, the kickbacks declined. Of course, that has little to do with how much of the billion dollars a year of fish moving through the market is contraband.

"We usually target specific dealerships with CIs [confidential informants]. Word goes fast when we show up," Scott explains. "We or state guys like Tommy will try and get down here two or three times a week. Of course, we could be here every night and still stay busy.

"Me and [NMFS agent] Jimmy MacDonald on Long Island try and cover the tristate area, including the three airports," he continues. "We're looking for black-market lobster, bluefin tuna [that can sell in Japan for $30 a pound], that sort of thing. We recently caught two fish at JFK. They were out of North Carolina and worth about $60,000."

Scott suddenly stops next to a puddle reflecting squiggly light from a nearby lamppost. "You ever seen *Guys and Dolls*?" he asks, grinning. "It's exciting like that with the wet streets, the fresh rainy air, the hi-lows unloading and all the characters around here."

I look around as we walk into the market. The streets are full of hard-working men and freshly killed fish—boxes, crates, handtrucks, and fork-lifts full of fish and half the guys carrying wooden handled metal hooks over their shoulders or in loops by their hips. There are open displays of big tile fish, king mackerel, and yellowtail flounder the size of serving trays, boxes marked Wisconsin farm sturgeon, farm-raised striped bass and Tilapia, along with baskets of clams and scallops, small red mullet, and yellowtail snappers in fading colors of red, yellow, sky blue, and green. A couple of guys cover crates of crabs with a blue tarp. There's fresh and frozen squid and Scott and Tommy and the dealers are acting friendly, greeting each other with tense nods and smiles.

"Want to keep it professional, never let it get personal," Scott mumbles. Tommy stops and checks the paperwork on some haddock and pollack. Scott pulls out a gauge to check some lobsters for size, measuring from the back of the eye to the beginning of the carapace. A short, old man in a torn cotton sweatshirt is introduced as Herbie. "How ya doin'? Nice to meet ya'. S'cuse me, but right now I've got a lot of stuff to do," he says in a gravelly voice.

"Herbie's company does $68 million in annual business," Scott tells me, as the old man wanders into a narrow cubicle marked M. Slavin & Sons. His company is one of five wholesalers charged in 1999 with buying illegally caught fish.

We pass 100-pound halibut from Maine on display outside other narrow streetside fish stalls, including that of M. V. Perretti Corp., another of the five accused wholesalers. In December 1998 a New Jersey fisherman, Ronald Ingold, pleaded guilty to illegally taking and selling fish from PCB-contaminated waters under the George Washington Bridge off Manhattan.

In court a year later, assistant U.S. attorney Joseph DeMarco claimed Perretti bought nearly 100,000 pounds of this tainted fish from Ingold and other fishermen.

"Everyone knows me here. If we want to do undercover work, we bring someone from outside," Scott explains. A bone-chilling cold has set in between the iced fish and drifts of rain. Forklift derbies are going on under FDR Drive. I ask a dealer about some odd-looking red fish. "They're silkies out of Brazil, a kind of snapper," he tells me. There are also piles of dead parrot fish, their colors, so dazzling when you see them munching coral among the runner reefs, are now faded to a dull algal green. "The West Indians buy them," the dealer tells me. "They go for $2.70 a pound."

DEC Lieutenant Tim Duffy joins us, a friendly, squared-jawed cop with erect posture and a clipped military-style mustache. We pass Pickle Barrel Seafood, a company recently assessed $80,000 for dealing in unre-ported fish. NOAA will later reduce this to a $2000 fine and a 10-year ban on trade in federal fish. A young guy pushes a handtruck past the DEC cops, singing "You must have been an ugly ass baby, cause baby look at you now." They smile coolly.

Big groupers are on display, possibly imports. All but the smallest groupers have been fished out of Florida. We pass a 50-pound bulbous-headed Louisiana buffalo fish, tagged clams, and Florida Jacks. There's a big tuna, a yellowtail out of Vietnam, and bunches of New Zealand clams and greenshell mussels. More snapper and small Atlantic swordfish, each maybe 80 pounds before their heads and tails were removed. Twenty years ago they averaged more than 260 pounds, but the longline fishing fleets, both U.S. and European, fished out all the big ones and began taking immature fish too young to breed.

We enter a shed-covered area called the Tin Building, where I spot a big swordfish, about 250 pounds. "That's the way they should all be," Scott says. I admire a big-eye tuna, which looks like a fat torpedo. The dealer has laid it flat on one fin. Since it was caught its skin has not touched a deck or floor. The Japanese buyers do not like bruised flesh. "This one's 158 pounds, worth about $550," the dealer tells me. "It's from Ecuador."

Tommy thinks he found some bags of undersized clams and pulls out a 1-inch gauge. If they slip through, they are illegal, but these prove to be legit, if just barely.

As the two state cops are checking the clams, Scott stands back and scans nearby stands to see if anyone is trying to move anything out of sight. A dealer strikes up a short conversation with him about the recent bombing of Kosovo. "That's what's called a distraction strategy," Scott says wryly after the guy moves on.

"See, they're fine," the dealer with the clams says with a grin. "You know me. I'm here 44 years."

"You're due." Scott replies, the hard cop.

Meanwhile, Tommy explains "stovepiping" to me.

"You have to dig through the bags 'cause they'll use a metal stovepipe to load undersized clams in the center, pull it out, and fill in all around with legal-sized clams."

Frank Giammarino, the owner of Seahorse, comes up to Scott to complain. "I got a $50,000 fine cause I didn't even know I needed a federal permit, but those boats in Montauk, they got $1500, $2000 fines. It's not fair. When I hear what they got, I'm upset. I want to appeal."

"Call me, I'll give you our attorney's number. The one who handled your case," the agent tells him.

"He had 14,000 pounds of summer flounder he shouldn't have had," Scott tells me after the dealer is placated, "from a boat that didn't exist."

On July 17, 1996, a Boeing 747 that had just taken off from New York's Kennedy Airport plunged into the ocean, killing all 230 people on board. The TWA Flight 800 disaster, initially believed to be the result of a terrorist bomb, led to a massive search and salvage operation in the waters off Long Island that continued for the better part of six months. Federal, state, and local agencies including the navy, Coast Guard, NOAA, and New York DEC were used in the search for clues and bodies.

During the search, with state officers off the docks and the Coast Guard not doing fisheries patrols, reports began filtering in of widespread pirate fishing taking place off Long Island and black-market product moving through the Fulton Fish Market. When Scott Doyle, Jimmy MacDonald, and other agents checked Seahorse's summer flounder records against paperwork in Montauk (on the tip of Long Island), they discovered a fictional boat named "Magpie" working with the unpermitted dealer. Several boats had been overfishing, including a 76-foot trawler named *Perception*, whose owner, Bill Grimm, and operator, Bill Braun, created "Magpie" as a cover for their taking of summer flounder out of season. Crates of these fish would appear on the dock in the middle of the night with a note from the "Magpie" crew instructing Seahorse to leave envelopes of cash pinned to a spot on the wall of the dockhouse.

Another dealer now comes up to Scott to complain how "the fishermen are going fucking crazy with all the rules and regulations. Could you tell me if it's okay to use fishtraps on tile fish right now or not? How are these guys supposed to know all that crap?"

"I couldn't tell you, but if you called me, I'd look it up and let you know. I'd never not get back to you," Scott promises, "and if I'm a tile fisherman

I will know the rules on my fish, 'cause that's my business, that's how I make my living, so I will."

"You got a point," the dealer concedes. "Who do you work for?" he asks me. I tell him I'm working on a book but also write for different outlets like *Sports Afield*.

"Oh, yeah, all those rich recreational fishermen who are trying to put us out of business. They're anticommercial fishermen."

"I also do stuff for Marketplace radio. You're not anticapitalist are you?"

"Fuck no, I'm a capitalist. I'm the best capitalist in the world!"

"Better than Herbie?" Scott wonders.

"I'm a capitalist, not a crook," he responds.

Tommy tells me how New York DEC cops like himself cover everything environmental, from fish on the waterfront to toxic waste to endangered species. They recently took a mountain lion out of a Brooklyn apartment.

"Where was it?"

"It was in the bedroom. We staked out the place after a tip and went in with Emergency Services. It was a full grown female. We tranked [tranquilized] her and now she's in a wildlife protection park somewhere upstate. We had another mountain lion sting where we bought a lion from a guy in the park. People seem to like the big cats."

We pass some Australian yellowtail, some octopus and live crabs, sea urchin roe and skate wings, catfish and grunts, whiting, and butterfish. Still, the Fulton Fish Market is tiny compared to others like Tokyo's Tsukiji. What they all have in common, though, is globalization, the creation of a world market for anything indigenous to the sea. The urchin caught in California, Maine, or Alaska one morning could have its gonads removed and served in a Tokyo nightspot the next evening. A bluefin tuna caught off Louisiana will definitely end up in Tsukiji, as will most black cod (sable) caught off the West Coast. A white abalone from California could be the centerpiece of a $450 dinner in Hong Kong, which is why there are only about 2000 of this extinction-bound species of sea snail left in the ocean. Giant geoduck clams caught in Puget Sound have been smuggled into Canada for shipment to Asia, just as polluted "black clams" have been smuggled from Mexico into the United States for sale in East Los Angeles. Things from the oceans once considered useless or inedible like baby eels, skates, dogfish, horseshoe crabs, and sea urchins all now have their markets.

The global fish trade is also keeping many Americans ignorant about what's happening in their own waters. People who order fish and chips in Boston may not realize that the white fish they eat is pollack from the

Bering Sea off Alaska instead of overfished New England cod. The blue crab ordered in Baltimore may come not from nearby Chesapeake Bay but Indonesia, where the pickers work for $15 a week. An expensive salmon served at a fine restaurant in New York or Los Angeles could be from a fish farm in Chile or Norway. Pacific Northwest salmon that once spawned by the millions are slowly going extinct—river by dammed, logged, and diverted river. Atlantic salmon that once returned to Maine's rivers by the hundreds of thousands recently have declined to about 30 adults. But when those salmon were proposed for inclusion on the federal endangered species list (which would restrict economic activities that threaten the fish), Governor Angus King called it a betrayal, arguing that Maine's native salmon can easily be replaced by farm fish from the state's booming aquaculture industry.

We pass by Joseph H. Carter, Inc., a dealership that three weeks later will be hit with a $1.72 million assessment for selling 1.25 million pounds of black-market fish, including endangered New England cod—although more than a year later NOAA will still be negotiating with the company to lower the fine. A Gloucester, Massachusetts, trawler, the St. Mary, will be seized and other trawlers fined.

By now it's 6 A.M. and the market's beginning to clear out. It won't be active again until after midnight. As we head back to our cars, Scott dumps some old coffee and a cardboard cup down a storm drain. Tommy and Tim give him a weird look.

"Of all the fish caught in the U.S., how much is being taken illegally, and how is that figure incorporated into the stock assessments the government uses when it decides if a species is overfished?" I ask our federal litterbug.

"The number of illegal fish is never going to be counted in the estimates," Scott replies. "Not with industry making all the rules."

It's a complaint I'll hear repeated often by enforcement agents, scientists, environmentalists, even a number of fishermen who recognize that the present system of industry-dominated management is broken, probably beyond repair, and that a new approach has to be taken if America's sealife is to survive.

"People say we're 40 percent overfished but we're 60 percent not," Andy Rosenberg, a top official at NMFS, argues when I suggest that the system's not working. Six months later his agency has raised the figure to 43 percent overfished. Along with the 230 commercial species whose fishing status NMFS is able to rate, they admit there are another 674 species whose status is unknown to them.

While managing fisheries one species at a time may help in tracking commercial landings, it provides no basis for understanding the complexity of saltwater ecosystems and how the removal of one type of marine animal can impact others. What is known is that, in the middle of a global extinction crisis, the world's aquatic species are going extinct at a rate five times faster than that of land animals.

Fisheries managers are also failing to account for the impacts of fish taken illegally or as bycatch (caught and thrown back over the side because they are not the fish being targeted).

In Puget Sound a few years ago investigators realized that the fisheries management plan on geoduck clams was underestimating the diver take by 100 percent. For every one of the giant bivalves harvested legally, another one was being pirated and shipped off to Asia via Canada. Local crabs were also being heavily targeted by poachers.

Pete Choerny, an NMFS agent, and I are on the cold, slate-green waters of Boundary Bay in northern Washington, below a bruise-colored, rain-saturated sky, passing through fields of crab pot buoys. "It's like a minefield trying to walk across this bay if you're a crab," says Pete, flashing a ready smile as he squints out across the water at three small fishing boats, lifting his binoculars for a closer look.

We're approaching latitude 49, the invisible line in the water that separates the United States from Canada. Craig Carlisle of the Washington Department of Fish and Wildlife checks landmarks along the coast against the GPS monitor on the bridge of the *Gauntlet,* a 46-foot state-owned cabin cruiser. "They're over a quarter mile south of the line," he notes, bringing the wheel about. "They're taking off like a covey of quail," Choerny agrees, as the three boats scatter, the *Gauntlet* going to full throttle to chase after the closest one, a 24-foot aluminum skiff with two men aboard. "This is a typical day on the line," Choerny half-shouts over the roar of the engine, as the powerful outboards on the skiff we're chasing throw up a rooster tail of water ahead of us. The skiff jags across the water to keep its stern to us, hiding the Canadian registration number on its side. After entering Canadian waters in hot pursuit but failing to gain on his quarry Carlisle powers down and turns his boat around. Choerny, Carlisle, and David James, a fisheries enforcement officer from the Lummi Indian reservation, begin pulling up the pirate traps south of the line.

"What would you have done if you'd caught them?" I ask. "Seize their boat and put them in jail," says Carlisle.

Knowing that the possibility of jail exists tends to make poachers and pirates uncooperative. Several years ago the *Gauntlet* was pursuing another illegal crab boat with another state officer and NMFS agent, Andy Cohen, on board.

"We came alongside during a chase in heavy 4-foot seas," recalls Cohen, a father of two with curly, dark hair. "I stepped on board and moved to turn off their engine. That's when one of them grabbed a meat cleaver off the transom. I pulled out my .357 revolver from my shoulder holster, backing up against the wheel. He moved forward, lifting the cleaver over his head. You know how with a revolver trigger there are two clicks before it fires. I had it pulled two clicks back. The state officer had his gun aimed from the other heaving boat, and I remember thinking if he fires it could hit any of us. The guy with the cleaver was ready to swing down on me when he decided to drop his weapon. He saved his life and saved me from ruining my life by his choice."

Along with poaching, bycatch is another fisheries impact that threatens the ocean's limited reserves of life. On a shrimp trawler on St. Simon Sound off Georgia, I watch three try net sets. The small try nets are used to test the waters before big commercial nets are deployed. The first set brings up a pile of white shrimp (that wholesales for $4 a pound) and a larger pile of bycatch, including whiting, silver eels, tonguefish, squid, butterfish, silver perch, and weakfish, plus a rare, spiny-looking sea robin and some bay anchovies. They are all separated out and shoveled back over the side. On the next dip we get about one-third shrimp along with much the same bycatch mix, plus some blue crabs, ocean catfish, baby flounder, and whelks. When this bycatch is swept off the deck the baby flounder, whiting, and tonguefish are grabbed off the surface by flocks of cawing seagulls that have descended on the boat like frat boys at a tailgate party. The third set brings up a big flounder that the crew keeps, a big stone crab, spot fish, pig fish, silver eels, tonguefish, and whiting.

In the Southeast shrimpers get 3 to 4 pounds of bycatch for every pound of shrimp they catch. In the gulf it's more like 7 to 9 pounds of bycatch per pound of shrimp. NMFS estimates that the U.S. shrimp fishery bycatch is close to a billion pounds a year, a waste of edible protein (unless you're a seagull) equal to 10 percent of the total U.S. catch. Drowned sea turtles also used to make up a significant part of this bycatch until NOAA required the shrimpers to add escape-hatch turtle excluder devices (TEDs) to their nets. Today the five endangered species of sea turtles found on America's Blue Frontier are making a modest comeback, while the shrimpers are making more money than ever. The use of TEDs was delayed for years, however, because of opposition by industry-friendly politicians, including President Bush's secretary of commerce (and NOAA boss) Robert Mosbacher, Representatives Tom DeLay of Texas and Billy Tauzin of Louisiana, and Representative (now senator) John Breaux of Louisiana, who got the House Appropriations Committee to cut funding for TEDs research.

Today many of the same shrimpers and politicians who opposed TEDs are fighting a proposal to develop and deploy BRDs (pronounced "birds"), or bycatch reduction devices.

As incredibly complex as the crisis of America's fisheries may seem, it can be broken down into three basic problems: the destruction of fish habitat, overcapitalization of the industry, and built-in conflict of interest in fisheries management.

"For years we've said, 'How can you address fisheries without addressing fish habitat?'" says Zeke Grader, director of the Pacific Coast Federation of Fishermen's Associations (PCFFA), which represents some 3000 small-boat operators on the West Coast. "The problem is you got the Farm Bureau and the oil industry saying NMFS has no business looking at essential fish habitat. And NMFS doesn't have the stomach to take on dam builders and developers in order to protect fish."

Along with the loss of rivers, wetlands, and productive coastal estuaries, recent studies identify certain kinds of fish farming and even fishing equipment itself as additional causes of habitat destruction.

Today 20 percent of the fish Americans consume comes from aquaculture, the farm rearing of aquatic plants and animals. Most catfish and trout, along with about half the shrimp and salmon we consume, are from farms, as is a large proportion of the talapia, striped bass, sturgeon, seaweeds, abalone, oysters, and crawfish. With consumer demand for fish growing and wild captures declining, U.S. aquaculture has taken up the slack, expanding some 6.5 percent annually. Right now 90 percent of U.S. domestic production comes from ponds in Mississippi, Louisiana, Idaho, and Appalachia. These aquafarms produce some waste water but have also been able to develop effective waste control systems. Offshore shellfish farming actually improves water quality, because the oysters, mussels, and other bivalves they raise pump up to 50 gallons a day through their shells, filtering excess nutrients out of the surrounding water.

It is the netpen farming of salmon in coastal bays and harbors that can damage marine habitat. High-density salmon farms pollute shoreline ecosystems with their excess feed and feces. Because the fish are so concentrated, they are also more susceptible to sea lice and diseases like infectious salmon anemia, which can spread to wild fish. But the antibiotics used to treat these farm fish can also prove fatal to surrounding aquatic life. In addition, Atlantic salmon used throughout the industry escape on a regular basis, including 100,000 that got loose in Washington's Puget Sound in June 1999. Washington's Pollution Control Board ruled that escaped Atlantic salmon meet the definition of a pollutant. The state of Alaska has gone further, banning all netpen aquaculture in state waters. The fear is that the escapees will spread disease, displace endangered wild Pacific salmon in their native rivers, or breed with them and weaken their genetic instincts for migration and spawning. There is also growing concern over companies trying to sell the industry's transgenic fish, genetically modified for fast growth (up to 150 pounds), and the possible impact of these "Frankenfish" when they escape into the wild.

Clearly, like many new industries, aquaculture holds both great promise and some peril, but where wild salmon and salmon fishermen are still viable indicator species for clean oceans, independent livelihoods, and healthy rivers, salmon farming doesn't make a lot of sense.

Nor perhaps does drag trawling, although this fishing method plays a much larger role in providing seafood for America and the world. About half the world's catch comes from bottom trawling gear for shrimp and fin-fish that use big-mouthed otter nets pulled along the seabed by chains and steel slabs called otter doors. Scallop and clam dredges are made of chain-mail bags and bar spreaders that also drag across the bottom behind plow-like metal hausers.

In December 1998 the scientific journal *Conservation Biology* published seven papers identifying bottom dragging as the major cause of ecological damage to the oceans' complex sea bottoms and the plants and animals that live there. Each year trawl nets scour benthic communities twice the size of the United States. The parts of New England's Georges Bank not shut down by overfishing are bottom trawled three to four times a year.

Peter Auster of the National Undersea Research Center at the University of Connecticut described putting on diving gear and sitting on the seafloor as a scallop dredge rumbled by. He then swam into the area to see what happened. He reported, "What was a complex of sponges, shells, and other organisms, was smoothed to a cobblestone street."

After Auster compared trawling to clear-cutting of the world's forests, the trawl studies received widespread (if brief) media attention. Fishermen, offended by the clear-cutting analogy, preferred one made by another study author, Elliott Norse of the Marine Conservation Biology Institute. Norse compared the effect of bottom trawling to land clearing on the last frontier that converted complex forest systems into simple cow pastures.

While many draggers like to think of themselves as farmers of the sea, until the 1980s there were extensive bottom areas of rocks, wrecks, and reefs that were inaccessible to their plows and chains. Then came new inventions called rockhoppers and street sweepers, big rubber rollers and brushes that allowed trawl nets to penetrate these last wild refuges of biodiversity without getting their gear hung up.

In response the state of Alaska banned bottom trawling in a number of areas to protect rocky crab habitat. Fish nurseries off North Carolina are protected from bottom gear, as are Florida's coral reefs and Maine's lobster coves. In California the state plans to expand a system of marine protected areas or "no-take zones" that will exclude not only bottom gear but all commercial and recreational fishing in an attempt to rebuild and propagate marine wildlife. The Magnuson-Stevens Sustainable Fisheries Act of 1996, the main federal fisheries law, requires protection of essential fish habitat. But by the end of the twentieth century no steps had been taken to enforce this provision of the law.

The size and extent of bottom trawling as a threat to habitat reflects an even bigger problem facing America's fishing industry—overcapitalization. For the U.S. fishing fleet that simply means there are too many (or too large) vessels catching too few fish. If the nation's 110,000 commercial fishing vessels were allowed to fish to their full capacity every day next year, there would be no fish left to catch the following year. Perhaps it might take the giant factory trawlers fishing in the Bering Sea an extra year or two to wipe out the pollack stocks, or gulf trawlers some additional time to kill off quick-breeding shrimp, but no one questions that the technical capacity to exterminate the Blue Frontier's wildlife is already tied up at America's docks. It is only through fishing seasons, closed areas, gear restrictions, licenses, and other regulations that the battered resource survives at all.

Ironically, this overcapacity came about in response to an effort by Congress to protect America's fishermen from foreign competition. In the 1950s New England's fishermen were astonished by what they ran into on Georges Bank. "They're fishing out there with ocean liners," they reported on returning to port. What they encountered were the first of what would grow into fleets of foreign factory-equipped freezer stern trawlers, known today simply as factory trawlers. These floating catcher-processor ships quickly began to outfish the smaller American boats. By the 1970s they were also stripping the offshore waters of most of their fish. American fishers from towns like Gloucester and New Bedford, Massachusetts, began appealing to the government for relief. The fact that a majority of the foreign factory trawlers were from Russia and Poland (along with Spain, Britain, and West Germany) helped provide a Cold War rationale for Congress's decision to ban the "spy-trawlers" and declare a 200-mile U.S. fishing zone.

The 1976 Fisheries and Conservation Act, also known as the Magnuson Act, after Washington Senator Warren G. Magnuson, who sponsored it, was not really about conservation. It was an assertion of exclusive U.S. fishing rights on the continental shelf, much as the Truman

Proclamation had been for oil. It also created a powerful precedent for the Reagan administration to follow when it declared America's EEZ Blue Frontier seven years later.

Among the Magnuson Act's provisions was the creation of eight regional fisheries councils to advise NMFS on how to promote U.S. fisheries, develop new fisheries, and establish maximum sustainable yields for fisheries. These maximum yields were supposed to represent the number of fish that could be taken without beginning to wipe out the stocks, although this theoretical number could be exceeded any year the councils determined there was an economic or social need to do so.

Because it was believed to be important that the councils include the expertise of professional fishermen, they were also exempted from conflict-of-interest laws that apply to every other federal regulatory body in America. What FEMA flood insurance had done for coastal protection and the Army Corps of Engineers for wetlands, the Magnuson Act now promised to do for America's fish.

"It was naive to believe you could get things to work like that," concedes Senator John Kerry, "but it was probably all that could be achieved at the time."

"We got rid of the foreign flag vessels that were raping our fisheries and then with our Yankee ingenuity figured out how to do it better," is how John Strong-Cevetich, a former Alaskan fisherman, explains it.

In the wake of Magnuson, the federal government created and expanded a range of fishing subsidies. Under the Capital Construction Program, fishermen could defer taxes on profits if they put the money into new boats. With the Fisheries Obligation Guarantee, or FOG, the United States pledged its full faith and credit against any loan for a vessel or fish processing plant. The government also set up the National Fish and Seafood Promotion Council, which advised consumers to "Eat Fish Twice a Week."

Just as many family farmers were encouraged to buy new combines and expand their acreage in the 1970s, fishermen in the 1980s were encouraged to take out multiple loans to upgrade their boats. Farm Credit Banks were among the major lenders in the gulf, while in Alaska, the Christiana Bank of Norway put $315 million into fleet expansion, including the construction of new American factory trawlers.

Fishermen, flush with easy credit, went on a buying spree, purchasing steel-hulled vessels with stronger engines, navy-developed Loran and fish-finding sonar, spotter planes, helicopters, and satellite relays that help locate fish congregations by tracking ocean surface temperatures.

There were dramatic production booms wherever this new fishing power was brought to bear on the resource, whether on reef fish in the Gulf of Mexico, the historic cod and scallop banks of New England, or the newly opened king crab waters of Alaska. Ports like Dutch Harbor in the Aleutians soon took on the look of nineteenth-century frontier mining towns with thousand-dollars-a-hand poker games, booze, speed, cocaine, and knife fights. Fishermen would rather slash each other than risk fist-fighting for fear of bruising their hands. If they couldn't haul crab pots, they might lose a $20,000 or $30,000 share on the next trip out. But trickle-down economics was about to take some of that money out of their hands anyway.

In 1986 the Reagan administration unveiled an investment tax credit that allowed people to take $10,000 off their taxes for every $100,000 they put into a new capital venture.

"I was trying to go from a 22-foot boat to a 35-foot fiberglass boat and I wanted to borrow $25,000 and I sent in an application and was denied cause I wasn't asking for enough money. You had to want at least $100,000," says Maine fisherman Paul Cohan. "The Reagan idea was let's give the big guys big investment credit. They gave incentives to all these doctors and lawyers to get into the industry."

"The biggest impact on the fishery was the tax changes Reagan made," agrees Andy Rosenberg of NMFS. "As a result you saw the fisheries grossly overcapitalized."

One of those grossly overcapitalized fisheries is in scallops, with some 300 boats now working New England (where the resource could probably support 100). I ask Rosenberg why, after NMFS agents recently seized five pirate scallop boats in Massachusetts, they were sold back into the fishery.

"When we seize boats we're directed to get the best return for the government," he explains. "Plus some of these boats have loans to pay off, government capital construction or FOG loans."

"I've busted poachers, real bad guys, and then gotten calls from higher-ups in NMFS," an enforcement agent confides. "They've told me, you can't put this guy out of business. He has government loans he has to pay off."

By the 1990s the capital-driven cycle of boom and bust had played out on the Blue Frontier, just as it did on the western frontier, where the railroads, hide markets, and Sharps repeating rifles allowed commercial buffalo hunting to reach an economy of scale.

King crab populations collapsed in Alaska; redfish, shark, and grouper in the Gulf; abalone and rockfish in California. In New England, where in 1784 the people of Massachusetts hung a golden cod in their statehouse to celebrate the abundance of their waters, large parts of Georges Bank had to be closed down to save the last remaining codfish. A number of Massachusetts trawlers quickly shifted their fishing power into the already depleted Gulf of Maine. Within four years that fishery had also collapsed.

I meet Rod Avila, a fourth-generation Massachusetts fisherman and owner of two trawlers, the *Trident* and *7 Seas,* at the New Bedford Fishermen's Family Assistance Center. He works there as an outreach specialist. It's a gray, rainy day. The walls of the waiting area are covered with cut-out paper fish with photos pinned to them. They are photos of young and not-so-young out-of-work fishermen who have "graduated" to new careers. The labels read "Brian Mayall, master mate (tugboat); Daniel Gray, trucker; Joseph Froias, computerized accounting; James Acmeida, computerized accounting; Derek Ealy, master mate (freighter); Alfredo Silva, truck driver."

"We've had 842 people come through the program," Avila tells me. "My idea is if they see a picture of a friend, they'll figure they can do it, too."

In an October 1992 editorial titled "Picking Up the Groundfish Tab," the *National Fisherman,* a major industry publication, wrote, "because of political ideologues in Congress who think raw competition, not cooperation, is the 'American Way' and even nature's way, our nation ran for the short-term money instead of a sustained, stable, resource industry. . . . When you run a resource-extraction industry without limits on capitalization, the workers—deckhands, skippers, and processing crew—not the capitalists, will pick up the tab in the end."

Not all that overcapacity was generated out of Washington, however. In south Florida the Colombian drug cartel lent a hand.

"I watched fisheries overcapitalized by drug smugglers," says Billy Causey, a one-time tropical fish collector and now manager of the Florida Keys National Marine Sanctuary. "A guy I knew went from one lobster boat to 10,000 lobster traps. Fishermen were getting $100,000 payoffs for one night's work. In the late 1970s and early 1980s these guys were getting rich and putting the money back into the only thing they knew, which was fishing."

During the early 1980s, NMFS agent Dan O'Brian worked the Keys and Florida's west coast. "I'd go alone into places like Everglades City and these guys couldn't believe I wanted to inspect their fish," he says, laughing. "I was always looking over my shoulder. The local sheriff [who was later arrested in a major drug raid] used to follow me around when I'd drive through there."

At one point O'Brian's informants told him that a couple of long-line fishing boats were going to pick up some Colombian hit men off the Dry Tortugas. They had been sent north to assassinate the U.S. attorney who had put drug boss Carlos Lehder in jail. O'Brian went out with a navy heli-copter, spotted the boats, and helped coordinate the subsequent Coast Guard boarding and arrest of several suspects. "Fishing and smuggling have always gone together," O'Brian claims. "I mean, who's going to notice when a fishing boat pulls into port?"

"If you're going to send armed terrorists aboard my boat, they're going to get an answer, 'cause I'm looking at death through defiance before I go out and put up with this foolishness any longer!" shouts Dave Marciano, staring hard at the Coast Guard representative to the New England Fishery Council, which is meeting in the basement of the Sheraton Plymouth. Behind the gold chain–wearing militant are 150 other pissed-off fishermen, some quite large and muscular from hauling nets and dredges. Four uni-formed Plymouth cops in the back of the room are trying to look unintimi-dated, while several more gather outside, talking to NMFS agent Dick Livingston, a former Secret Service agent who really is unfazed, having seen this all before. Reporters and camera crews from the Boston TV stations have also come out for the show.

After Dave makes a few more threats, Paul Cohan gets up to play the voice of moderation. Paul has a reddish-gray ZZ Top beard and a white billcap.

"If we can manage to see some motion toward us and a little bit of com-passion, that will help, help the fish, help us help the fish by letting us land what we're catching. We can all come out of here having accomplished something today," he promises. "But if the Fisheries Service is going to hold the company line and remain inflexible, what will happen is that every man in this room and everybody that's not here is just going to go fishing, and we're going to bring in what we catch and the hell with the closed areas, the hell with the days at sea, because we can rip up citations just as fast as you can write them, and then you're going to have a political crisis that's unparalleled."

A number of fishermen get up to testify that there are ten times more cod in the Gulf of Maine than NMFS biologists are claiming. They can hardly catch a flounder without some cod bumping it out of the way to climb into their nets.

"The *Albatross* [NMFS research trawler] couldn't catch a fish if they towed in the New England Aquarium," Massachusetts fisherman Paul Terrio declares to hearty applause.

"We're starvin' here," a 30-something fisherman yells out. Later, I spot him climbing into a late 1990s extended-cab truck in the parking lot.

The issue of the moment is cod discards, but the overall issue is how the fishing industry controls the councils and how NMFS acts as its codependent partner in a dysfunctional relationship that would embarrass Jerry Springer.

By 1999, after five years of Georges Bank closures, cod were slowly beginning to reestablish themselves in the no-take zones. At the same time, with prices high, there was increased pressure to target the fish, even if they still represented only a tiny fraction of their historic abundance.

Similar closures were established in the Gulf of Maine, where the cod are near all-time lows. At the previous council meeting in January the fishermen promised not to target cod in the Gulf of Maine if the council would give up on a plan to expand the closed areas. To prevent a targeted fishery, NMFS said fishermen could only keep 400 pounds of cod bycatch a day. Suspecting the fish were still being targeted, NMFS quickly cut that to 200 pounds and finally to 30 pounds, about the weight of two fish. Any additional bycatch has to be thrown overboard. Which is why these fishermen are outraged.

"We can't mindlessly discard the very fish that we're trying to save. It serves no purpose. It doesn't do the resource any good, it doesn't do the consumer any good, it certainly doesn't help the harvesters, and it makes you guys look like jackasses," Cohan argues.

"My kids think I'm nuts, but I'm not nuts, you're nuts," another fisherman tells the council.

After more than six hours of abuse, directed mainly at the NMFS and its scientists, a couple of fishermen members of the council offer up a motion requesting the secretary of commerce to declare a resource emergency, allowing the fishermen to keep at least 700 pounds of cod per day.

The council chairman Joe Brancaleone, an ex-fisherman and now an executive with Burger King, suggests that "we need this type of motion or another closure."

"Or war!" someone shouts, not liking that closure reference.

Joe calls the question.

As tension mounts in the room, the vote goes down to the wire, eight for, eight against. Joe casts the tie-breaking vote for the motion. The fishermen are appeased.

"What you saw isn't typical," one of the council members tells me after the near-riotous meeting. "This only happens maybe two or three times a year."

Later, I talk to a reporter friend who has been covering the New England fisheries crisis for years. "In terms of the story, it might have been more interesting if the vote had gone the other way," I admit. "But the vote never goes the other way, and that's the real story," she points out.

After the vote, NMFS decides (against the advice of its own biologists) to return the take-home catch to 400 pounds a day. As the availability of the fish has declined, the price has gone up. Cod, which once wholesaled for 50 cents a pound, is now going for $2.50, which means that while licensed to go after other species, fishermen in the Gulf of Maine can now make $1000 a day catching what's supposed to be a protected species of fish.

It's a lose-lose scenario in which the market's increasing prices on diminishing stocks creates a disincentive for conservation, so that the last fish in the sea should be worth a fortune.

"There were some of us who should have spoken out yesterday, but we were intimidated both physically and emotionally," John Williamson, a retired Maine fisherman tells me the day after the vote. He was one of the eight council members against the motion. "The manipulation of the system is outrageous," he says. "I didn't sleep all night thinking we should have spoken out and maybe swayed that one vote. As Maine fisheries collapse, you see power shifting to Massachusetts. Those guys crying yesterday how they were starving. They had their best year ever last year and that's a fact."

Along with six voting members from federal and state fisheries agencies, the New England Council has 11 voting members appointed by the secretary of commerce. Eight of them are former or current commercial fishermen or represent commercial fishing interests. Two are from the sports fishing charter business. The eleventh, Doug Hopkins, is a lawyer with Environmental Defense and the only environmentalist on any of the nation's eight regional councils. Some councils however, such as the Mid-Atlantic and South Atlantic councils, are more heavily weighted toward recreational fishing. These "rec-fishermen" represent another multibillion-dollar user-group, increasingly in conflict with the commercial industry over allocation of the resource.

The political allocation of seats on the Fisheries Councils is yet another bizarre phenomenon worthy of a major political science, or perhaps anthropology, paper.

A few weeks after the New England Council meeting, I attend a U.S. Senate hearing on the state of America's coral reefs. Senators Kerry, Stevens, and Breaux are there, along with Senator Olympia Snowe of Maine, who chairs the Oceans and Fisheries Subcommittee of the Senate Commerce Committee. After testimony from the Cousteau Society, Waikiki Aquarium, and others about the declining state of America's coral reefs, Snowe, dressed in power pink and pearls, shows the first strong emotion of the afternoon when she addresses witness Sally Yozell, deputy director of NOAA. But it is not the dying of America's jewel-like living reefs that has sparked Snowe's ire. She's fuming that the secretary of commerce has named someone from New Hampshire to one of two New England Council seats traditionally reserved for Maine (he's actually from Maine but fishes out of New Hampshire).

"I want to make sure everyone in your agency understands how unhappy I am and the industry itself is and to have lost those seats [sic] to New Hampshire," she snaps. "We didn't even have a chance to discuss this with

Secretary Daley, although calls were placed to his office. . . . So I just want
to make sure that you will express my dissatisfaction to the secretary."

"I'll let him know how you feel," Yozell responds meekly.

Snowe later passed a rider to the Magnuson Act that added a twelfth
appointee (from Maine) to the New England Council.

"It's all too political," says Rod Avila, who served as a member of the
New England Council for three years before quitting. "I saw too much of
that self-dealing on the council, too much people going out to lunch to trade
their votes and making backdoor deals. I think it should be all government
controlled to do what's best for the resource."

Unfortunately, the government agency responsible for protecting
America's living marine resources has failed to do its job.

"NMFS is charged with both conservation and promotion of seafood
consumption, but NMFS is also located within Commerce where its commercial function dominates," explains Representative Jim Saxton of New
Jersey who has chaired the Oceans Subcommittee of the House Resources
Committee. "Let me tell you a story," he continues. "A few years ago my
[district's] commercial fishermen came to me and said: 'We're in the scallop fishery and there's not enough scallops. There's too much [fishing]
pressure. We suggest you do a federal buyout, and we'll be happy to find
owners to retire their boats, turn them into reefs, sink them, or whatever
and those of us who want to stay in the fishery for the long term, we'll help
out [financially].'

"I was excited by these conservation-minded scallopers, and the
Virginia guys felt the same way as the New Jersey guys, but the New
England guys felt differently. They felt like, 'If we can open Georges Bank
back up, we won't need a buyout,' and you know what? NMFS went along
with them. We had things moving swiftly until they reopened the Georges
Bank to those scallopers. Now there's little incentive for a buyout."

In 1996 the Magnuson Act was reformed with lots of environmental
conservation language added on about protecting fish habitat and reducing
bycatch. There was even a provision saying council members should not
vote on fishing matters that would give them a "substantially disproportionate benefit" over other fishermen going after the same fish. The basic
conflict-of-interest exemptions were maintained, however, guaranteeing
that "active participants" in affected fisheries would be appointed to each
council. Rather than ban Jesse James from the railroad commission, these
rules ensured that the robbers divided the loot evenly.

The new law was named the Magnuson-Stevens Act in honor of
Alaska's Senator Ted Stevens, who promoted it. If the Democrats had controlled Capitol Hill, it probably would have been named the Magnuson-

Studds Act, after Massachusetts Representative Gerry Studds, who for years chaired the Merchant Marine and Fisheries Committee and was the driving force behind the House version of the reauthorization bill.

Part of the Magnuson-Stevens Sustainable Fisheries Act says that the secretary of commerce can declare an emergency and take unilateral action to protect living marine resources when a regional council fails to act. Despite fishing activities that threatened the survival of marine mammals, seabirds, and numerous species of fish, the Clinton administration's Secretary of Commerce William M. Daley chose not to take any action until 1999, when he wrote a letter to the New England Council telling them to open up closed areas of Georges Bank to scallop dredging.

After a scientist reported a big buildup of scallops in the closed area, the scallopers in Fairhaven (across the harbor from New Bedford) set up a group called the Fisheries Survival Fund. With scallop meats selling at more than $5 a pound, they decided to assess themselves up to $500 per boat ($45,000 a month). "We figured we could hurry up the process this way," explains the fund's Bob Bruno. With the money they hired a well-connected Washington lawyer to lobby for their cause and the recently retired Representative Studds as their champion.

Studds got Senator Ted Kennedy and Representative Barney Frank to meet with Daley to encourage the secretary to do right by the scallopers, despite New Bedford fishermen's complaints at council meetings that dredging would produce an unacceptable bycatch of protected groundfish (Fairhaven is a scallop port, New Bedford a groundfish trawler port). In fact, the first experimental dredging in a closed area brought up two pounds of small yellowtail flounder for every pound of scallops.

Massachusetts Senator Kerry chose not to join his fellow Democrats— Kennedy, Frank, and Studds—in strong-arming the secretary of commerce. "I don't condemn them but I thought what they did undermined the process a little bit," he says rather diplomatically.

In early 2000, a year after intervening for the scallopers, Daley again issued an emergency decree, this time delaying implementation of a conservation plan, put forward by the Mid-Atlantic Fishery Council, that some Gloucester and New Bedford fishermen (and their congressional delegation) did not like. It would have stopped the targeted fishing of large female dogfish, a slow-breeding species of shark that takes up to 20 years to reach sexual maturity. The British pay Gloucester fishermen about 15 cents a pound for the large females, which they then put into their fish and chips (traditional cod no longer being available).

While the New England Council management process is perhaps more colorful than some others, similar failures to protect the resource go on in all the councils, state fisheries programs, and among NMFS managers. The effect is being felt both in the water and on the docks.

I'm visiting the groundfish trawlers that are giving way to weekend pleasure craft in the scenic coastal town of Bodega Bay, California. This is where Alfred Hitchcock filmed *The Birds,* and the cawing gulls on the roofline of the Eureka Fisheries warehouse are doing their best to maintain a fearsome reputation. I watch as a trawler pumps thousands of pounds of black cod, orange-colored thorny heads, sole, and various species of sharp-finned rockfish into large plastic crates called totes. The company's weigh master records the fish poundage as the totes are forklifted onto a scale and then onto refrigerator trucks for the two-hour drive south to San Francisco.

Down on the Bodega Bay municipal pier, second-generation fisherman Andy Philips is directing his two-man crew, fixing the net on his trawler, the *Jo Ellen,* before heading back out to sea. A big man, with a gray Mennonite beard, black T-shirt, jeans, and boots, Philips has a swagger from 40 years on the ocean and a voice rough as salmon gravel. "The real problem is the regulations aren't doing the job," he claims. "Back around 1986 you'd just head out, and when the boat was full the trip was over."

"But you agree the fishing's gotten worse?" I ask.

"Of course the fishing's worse. We killed them all. Electronics [fish-finding sonar] killed them all. The fishing grounds are like freeways now and we're just wiping them out. Places where I used to get 20,000-pound drags I get 100 pounds now. Decimated is the word. If we just stopped fishing for about 20 years the fish'd come back. But that's not practical. I couldn't do any other kind of work myself."

"Who hears the fishes when they cry?" wondered poet Henry David Thoreau. Sylvia Earle does, I'd venture. The former chief scientist for NOAA (under President Bush) and famed ocean explorer, labeled "the country's foremost oceanographer" by NBC News and a "Hero of the Earth" by *Time* magazine, believes that commercial fishing no longer makes sense.

"It's not a harvest. It's the commercial taking of wildlife and there's no history of this ever having been done sustainably," she argues. "The idea of continuing to take hundreds of millions of tons of wildlife is inexcusable, and with these bottom trawl nets! I use the analogy of taking squirrels and rabbits out of the forest using a bulldozer. Unfortunately, the Bureau of Commercial Fisheries was reborn as the National Marine Fisheries Service not to serve the fish but the fisheries industry. I think if people knew fish not as fish but as these amazing animals I've gotten to know, things might change."

"Change has to come, provided we can get beyond this bigger is better mentality that was with us through much of the twentieth century," agrees

Zeke Grader of the Pacific Coast Federation of Fishermen's Associations (PCFFA).

PCFFA, along with the Maryland Watermen's Association, Cape Cod Commercial Hook fishermen, and a few other community-based fishing groups, believes sustainable fishing is still possible.

"For food production, smaller units make sense, plus the cultural element is involved," Grader argues. "Instead of indiscriminate trawlers we need new technologies for gear that's more selective, that harvests less but gets more value. Right now sardines are coming back in California, so we have to learn not to repeat our mistakes of the past, learn to fish them sustainably. Also what if instead of 40 to 70 boats fishing 50 to 100 tons a night and grinding them up for fish meal, we had 1000 boats fishing a ton a night and going into the fresh food market? Instead of surimi [factory-trawler blocks of processed pollack], the Velveeta of fish with the nutrients all washed out, what if we provided pollack as a low-priced white fish fillet for the supermarket? That way it wouldn't always be so expensive to buy fish, and lower-income folks could afford it also."

For the past 40 years the PCFFA has been working in coalition with recreational fishermen and environmental groups including the Sierra Club to find innovative approaches to the use and protection of local fish and marine habitat. They originally came together to fight water diversions that threatened California's salmon rivers. As a result of their early and ongoing collaborative work, California today has viable commercial stocks of wild salmon.

"These groups that are fighting each other elsewhere are missing the big picture," Grader believes.

A practical solution to America's fisheries crisis has to be at the heart of any big-picture approach to America's Blue Frontier. What is required is a public understanding and commitment to turning things around. That could be done using a combination of already available policy tools.

Let's call this solution the BLUE plate special. The BLUE, of course, is yet another fisheries acronym, but an easy one to remember.

The B is for buybacks, a financial commitment by both government and industry to reduce the size of the fishing fleet to a sustainable level.

"But why should the taxpayer buy back boats we may already have helped pay for?" I ask Senator Kerry.

"This country has historically helped people hit by sudden dislocation with retraining and other support, besides which there's no other way to reduce the fishing pressure. It's an effective approach, but only if done in combination with good management and closures."

Which goes along with the L in BLUE. Limited entry means only so many people can be licensed to work in a given fishery or biological complex of fisheries in order to prevent them from being overcapitalized again. Some people like the idea of privatizing the fisheries with individual transferable quotas, or ITQs, in which a fishing license is like an ownership deed to a given share of the fish stock. Others worry that this method will encourage corporate consolidation. In some places ITQs might work, in others not. Rather than get hung up on a single tool, it is important to stick with the larger principal of not allowing more people to fish a living resource than its biology and habitat can sustain—thus limited entry.

The U in BLUE is for undersea reserves, or what are being called marine protected areas. Biologists suggest 20 percent of the Blue Frontier needs to be set aside as limited or no-take zones in order to restore and propagate new populations of fish, crustaceans, and other plants and animals. Where undersea reserves already exist, new studies are finding them highly effective, with healthy populations of marine wildlife slowly expanding beyond their fluid borders.

Finally, the E in BLUE is an end to conflict of interest. Fisheries management must be taken away from people with a direct stake in killing the resource. At a hearing in Washington on the billion-dollar-a-year pollack fishery's impact on Steller sea lions, I heard a one-time NMFS scientist give testimony. He had quit NMFS to help found the Arctic Storm factory-trawler company and was also vice-chairman of the Pacific Fisheries Council. Such a direct conflict of interest does not exist in other industries. For example, an FAA inspector might quit his or her job to found an airline company, but once in that position that airline executive would not be allowed to sit on the National Transportation Safety Board. The flying public would not tolerate it, nor does the law allow it. The same principal of not letting economic self-interest oversee the public trust should apply to preserving our living oceans for future generations.

Still, a program involving buybacks, limited entry, undersea reserves, and an end to conflict of interest in our fisheries is not likely to take place until far more Americans who say they love the oceans decide to take more responsibility for their stewardship.

Unfortunately, under the present system of ocean governance there is no real stewardship or governance. Instead of seeing the Blue Frontier as a living entity, we have encouraged an array of special interests to attach themselves to various calcified bureaucracies, like so many poisonous anemones clinging to hard rock corals. That leaves the majority of us to play the role of passing small prey caught in their tentacles, injected with venom, and slowly drowning in red tape.

Chapter Ten

Drowning in Red Tape

No one knows who's on first, or even if they're playing baseball.

—KATHY METCALF, CHAMBER OF SHIPPING OF AMERICA,
ON U.S. OCEANS POLICY

We all drink bottled water, the streets are crowded, the beaches are closed and the reefs are dying. You tell me who's in charge?

—KEY WEST "REEF RELIEF" ACTIVIST DEEVON QUIROLO

Among the sins of Richard Nixon few historians count more than 30 years of failed U.S. oceans policy. Perhaps they should. On July 9, 1970, the same day he established the Environmental Protection Agency as an independent arm of the government, he created and buried another entity, the National Oceanic and Atmospheric Administration, assigning it to the Department of Commerce, then being run by his campaign fund-raiser and future Watergate bagman Maurice Stans.

"NOAA wasn't quite stillborn, but it was born feeble. Nixon did the minimum he had to," charges Edward Wenk, former White House secretary of the National Council on Marine Resources under both Lyndon Johnson and Nixon. Representative John Dingell of Michigan blasted the president's action, describing the newly established NOAA as the handmaiden of a Department of Commerce so dominated by industrial interests, "as to be incapable of objectivity on issues of the marine environment."

Logically, an agency designed to study the weather and protect the nation's oceans might have found a home in the Department of the Interior, whose job is to manage and protect America's public lands and wilderness. The smart money in the marine community certainly believed that if NOAA was not going to be an independent agency, then the Department of the Interior was where it would find a home.

What the smart money failed to realize is how personal spite and vindictiveness can have hugely disproportionate effects on public policy inside the Washington Beltway.

A few months earlier, on April 30, 1970, Nixon had ordered U.S. troops in Vietnam to invade neighboring Cambodia, which set off campus protests and National Guard and police killings of four students at Kent State, two at Jackson State, and one at the University of California at Santa Barbara. Deeply disturbed by this turn of events, Secretary of the Interior Walter J. Hickel wrote a personal letter to the president. In it he expressed his growing reservations about Nixon's refusal to listen to the antiwar sentiments of the nation's young. Hickel's letter, dated May 6, 1970, read in part, "About 200 years ago there was emerging a great nation in the British Empire, and it found itself with a colony in violent protest by its youth—men such as Patrick Henry, Thomas Jefferson, Madison, and Monroe, to name a few. Their protests fell on deaf ears, and finally led to war. The outcome is history. My point is, if we read history, it clearly shows that youth in its protest must be heard."

Before reaching the White House, a copy of the letter (which had been circulated at Interior) was obtained by the Associated Press and published in the *Washington Evening Star*. The president and his aides Bob Haldeman and John Ehrlichman went ballistic. Nixon told Hickel he now considered him an "adversary." Hickel was blacklisted from White House events and became the target of a well-orchestrated campaign of press smears. Less than two months later NOAA was placed with the ever-loyal Maurice Stans at Commerce. On Thanksgiving eve, Hickel was fired.

It was a rather sad outcome to one of the more hopeful initiatives of the 1960s, but one that went largely unnoticed in the polarizing political heat of the times.

For a period in the 1960s exploration of "inner space" was seen to be at least as important as work in outer space, with astronauts and aquanauts (including astronaut-turned-aquanaut Scott Carpenter) competing for national news coverage. Groups of navy and civilian scientists were living in underwater "habitats," like Sealabs One and Two off San Diego and Textite off St. John in the U.S. Virgin Islands. Major corporations including GM, Union Carbide, Lockheed, Reynolds, and Alcoa competed for what they imagined would be multibillion-dollar contracts if ocean exploration went the way of the space race with the Soviet Union. Advocates for new ocean spending were also not averse to citing the Red menace as justification for America's getting wet. In 1959, two years after the Soviets launched *Sputnik*, the first human-constructed satellite, Senator Warren G. Magnuson of Washington declared, "Soviet Russia is winning the struggle for the oceans. Soviet Russia aspires to command the oceans and has mapped a shrewdly conceived plan, using science as a weapon to win her that supremacy." And six years later, in 1965, he wrote, "The prevention of communist domination of the seas is perhaps our most pressing problem today. . . . This is the immediate challenge our marine scientists can and must help us meet."

New submersibles like the navy's *Alvin* were launched into the depths, and the popular imagination fired up by salty media ranging from Arthur Clarke's science-fiction novel *Deep Range* and Cold War potboilers like *Ice Station Zebra* to television programs such as *Flipper* and *Sea Hunt,* the latter starring Lloyd Bridges as underwater diver and investigator Mike Nelson. There were Jacques Cousteau's books, films, and lyrical "Undersea World" TV specials, as well as the pop sounds of the Beach Boys, Jan and Dean, the Ventures, Dick Dale, and many others who redefined the California dream as a surf safari along the golden shore of youth looking for that perfect tubular wave.

Vice President Hubert Humphrey, who as a senator had issued a 1957 decree on the importance of oceanographic studies, became a major advocate for new approaches to America's Blue Frontier. Of course, many of these approaches, techno-optimistic visions conceived in the 1960s, appear strangely anachronistic by today's standards. There was talk of developing a fish protein concentrate made up of whole fish, that could be added to rice, milkshakes, and other products to feed the hungry majority of the world's three billion people. Senator Claiborne Pell of Rhode Island, in his book *Challenge of the Seven Seas,* imagined that by 1996 there would be nuclear-powered underwater vacation resorts, submarine oil tankers tapping sub-Arctic oil fields, and surgically altered "fishmen" respirating through artificial gills.

In 1964 the White House Office of Science and Technology proposed building nuclear power plants all along the coasts to desalinate seawater for America's thirsty cities, beginning with Key West, Florida. Later there was a proposal to build nuclear power plants on a series of artificial islands off the coast of New Jersey. The Rand Corporation, Scripps Institution, and others looked to Antarctica as a source of freshwater. The National Science Foundation even drew up a proposal for a 20-mile-long iceberg convoy to bring frozen water north to California. The lead berg would be equipped with ships' engines and propellers. Today, a more modest proposal calls for using giant bladders and single-hull tankers to bring freshwater south from British Columbia.

Beginning in 1966 Vice President Humphrey took charge of the White House Council on Marine Resources and helped launch the Stratton Commission, a blue-ribbon panel convened to consider America's future in the sea. The commission was headed up by Julius Stratton, president emeritus of the Massachusetts Institute of Technology and board chairman of the Ford Foundation. Also on the 15-member commission was attorney Leon Jaworski, who would later be appointed Watergate special prosecutor assigned to investigate the Nixon White House.

In 1969 the Stratton Commission issued their report, "Our Nation and the Sea." Its ideas resulted in the passage of a number of ocean protection laws including the Coastal Zone Management Act and Marine Mammal Protection Act. But their key recommendation was for the creation of a unified ocean agency responsible for the stewardship and exploration of the Blue Frontier. It should be an independent agency, they proposed, and include the Coast Guard, Bureau of Commercial Fisheries, and National Weather Bureau. Suggested names for the new organization included Sea Exploration Agency (SEA), National Marine Agency, and National Oceanic and Atmospheric Agency. The commission envisioned a watery twin to the starbound NASA (*Science* magazine even christened it "a wet NASA").

In the fall of 1968 Secretary of Transportation Alan Boyd got wind of the proposed agency. He was outraged that anyone would suggest removing the Coast Guard from his two-year-old department, which had just wrested it from Treasury. Boyd complained to President Johnson, who considered the Department of Transportation his personal brainchild. Not only was the idea of moving the Coast Guard deep-sixed, but Johnson refused to meet with Stratton. Humphrey, who had championed the commission, felt humiliated by this slight. He experienced a far greater humiliation as the 1968 Democratic nominee for president (following Johnson's decision not to seek reelection), when he was unable to make a clean break from the domineering Texan's Vietnam War policy. Humphrey lost a close election to Richard Nixon, who promised the nation he had a "secret plan" to end the war.

After his inauguration President Nixon, like his predecessor, quickly became obsessed with winning the intractable war in Southeast Asia. When it came to creating and nurturing a new agency for the oceans, however, he proved more inclined to sink it like some Vietnamese sampan.

And so NOAA came into being in the relative obscurity of the trade-oriented Department of Commerce. With no strong advocate in the White House or Congress, its first budget of $330 million was $120 million less than requested, and almost all of that was dedicated to NOAA's "dry side," the Weather Service.

Robert White, director of the Weather Service, also became NOAA's first director, a job he held for seven years. "For awhile I wondered if the word NOAA would even take hold," he recalls. "We were a collection of all the cats and dogs of the ocean community, all the programs that weren't strongly attached to their parent agencies. So people still continued to refer to the Weather Service, or the Coast and Geodetic Survey, or the Bureau of Fisheries. The navy sent us a three-star admiral as a liaison, and he was very helpful. The navy of course wanted to keep closely associated with anything having to do with the oceans."

In the three decades since its founding, NOAA's various directors, designated undersecretaries of commerce, have come from navy-linked oceanographic institutes like the University of Washington and University of Rhode Island and continued to support a strong emphasis on basic research as a means to "advance our knowledge of the oceans." This science orientation has allowed them to avoid making the hard choices often associated with natural resource management.

Science, after all, is about presenting, challenging, and refining hypotheses over long periods of time to better understand how things work. Policy, by contrast, is about taking the best available science and, based on society's shared values, making decisions—decisions that are often controversial and result in winners and losers. By emphasizing the uncertainty of science and the need for more study, NOAA's directors have avoided making policy decisions that might impact business interests who look to the Department of Commerce for support and not regulation.

For example, take the salmon (which almost everyone has). A salmon migrates up the Columbia and Snake rivers bringing nutrients 900 miles from the ocean to enrich the granite soils of Idaho. There it spawns and dies in its natal river gravel or has its bones and skin deposited in the forest by some fur-bearing predator. As its young progeny heads back downriver and out to sea, it will have to make it through some three dozen different governmental jurisdictions that are all influenced by the votes and money salmon lack—by ranchers, loggers, hydroelectric dam operators, and their beer- and soda-can producing customers in the aluminum industry, by shoreside developers and the International Association of Shopping Centers, and the fishermen who all want a piece of that fish or its habitat. Unfortunately, over 30 years of declining Northwest salmon stocks, NOAA managed to ignore and delay an effective response until an environmental lawsuit under the Endangered Species Act forced the creation of a state and federal task force in the late 1990s. That task force has now

begun spending more than $100 million a year and has become embroiled
in a debate over dam removals, in a desperate eleventh-hour attempt to save
this living icon of Northwest wilderness.

Today, with no strong national leadership in sight, responsibility for
our Blue Frontier remains up for grabs, claimed by more than half the president's cabinet departments, at least 15 federal agencies, 44 committees and
subcommittees of Congress, and hundreds of state and local authorities
from 22 coastal states, the Commonwealth of Puerto Rico, and various U.S.
territories including Guam and American Samoa.

The following is a simplified breakdown of how America's oceans are
presently managed:

NOAA, on its wet side, remains responsible for federal marine science,
marine sanctuaries, fisheries management (beyond state waters), and
coastal management (through state agencies). The Coastal Barrier Act,
however, is overseen by

U.S. Fish and Wildlife Service (for reasons no one at FWS is quite sure
of). The service also protects marine mammals like walruses, manatees,
and sea otters while

National Marine Fisheries Service (with input from the federal Marine
Mammal Commission) is responsible for seals, dolphins, and whales.
NMFS protects sea turtles at sea while the Fish and Wildlife Service
protects them on the beach. If something happens to a turtle in the surf,
it become a jurisdictional dispute.

Department of Agriculture oversees the care of captive dolphins and
promotes aquaculture (as does the U.S. SeaGrant Program).

National Park Service takes care of dozens of "ocean units" including
national seashores and underwater areas adjacent to national parks.

National Oceanographic and Atmospheric Administration oversees
national marine sanctuaries.

Mineral Management Service (MMS) leases oil, gas, and mining rights
on the continental shelf (but not in state waters). On occasion MMS's
activities are sidetracked when oil exploration poses security problems
for the navy. MMS also faces periodic suits (under the Coastal Zone
Management Act) from California, North Carolina, and other states
that do not want oil platforms anywhere near their coasts.

Army Corps of Engineers is responsible for protecting coastal wetlands
(and handing out permits for their destruction) along with the EPA.

Environmental Protection Agency runs the National Estuary Program, not to be confused with

NOAA's National Estuary and Research Reserves, or the

Department of Agriculture's Natural Resource Conservation Service, which is working to restore wetlands in Louisiana.

In addition,

Army Corps of Engineers is charged with protecting the shore against flooding and sea surge and keeping coastal traffic moving through construction of seawalls, beach replenishment, and dredging of ports and canals.

EPA regulates the corps dredgespoils if they are toxic and oversees ocean dump sites for toxic and nontoxic muds. EPA is also responsible for regulating polluted coastal runoff (under the Clean Water Act), as is

NOAA (under the Coastal Zone Management Act).

Federal Emergency Management Agency (FEMA) steps in when the Corps of Engineers fails to prevent hurricane damage in high-risk coastal zones with its half-trillion dollars of flood insurance, allowing people to rebuild in harm's way.

U.S. Coast Guard, aside from search-and-rescue and law enforcement, manages ship traffic; controls ship, port, and marina-based pollution; and is responsible for oil spill response (under OPA 90), fisheries enforcement (with NMFS), and keeping track of northern right whales so ships don't run over them.

Department of Transportation (DOT), which oversees the Coast Guard, also licenses U.S. ships through the Maritime Administration (with 47 ships subsidized by the navy as a military reserve fleet). The DOT also builds roads, bridges, highways, and other infrastructures along the coast, often through wetlands and to and from barrier islands.

Department of the Interior oversees the 200-mile EEZs of U.S. protectorates like Midway and the northern Marianas through its Department of Insular Affairs.

EPA maintains clean-water agreements with former U.S. protectorates including Palau and the Federated States of Micronesia.

State Department negotiates EEZ rules, coral reef protection, and fishery treaties with other nations and, under a 1995 National Security

Council finding, considers itself the lead agency for all U.S. ocean activities. Since the finding was semisecret, no one else seems to know this.

National Science Foundation funds ocean research on climate change and other marine issues, as does the navy, NOAA, and NASA.

NASA also studies the oceans from space and is recruiting oceanographers to go into space. Similar networks exist within various coastal regions, states, counties, and municipalities.

Is that clear? Of course not. Unfortunately, this murky mess of competing bureaucracies tends to administer marine activities with little or no regard to their natural interaction within the oceans or the watersheds that flow into them. They are frequently criticized for failing to communicate with each other, failing to listen to their own scientists, and failing to solicit input from conservationists, coastal citizens, and communities.

It has been suggested by some in Congress and the White House that NOAA be given greater leadership responsibility for effective oversight of America's Blue Frontier. Then again it has been suggested by some in Congress and the White House that the FBI be given backdoor access and encryption keys to every home computer in America. Despite rapid growth during the 1990s, including important work on climate and an expansion of marine funding, NOAA seems to have lost the trust of too many people working on the Blue Frontier to maintain its credibility.

"There should be a separate agency for the oceans, or at least not one in constant conflict," argues former NOAA chief scientist and *National Geographic* explorer-in-residence Sylvia Earle. "The leadership has not been strong on biology, but biology is coming back because of this great wake-up call from nature."

When Roger McManus, the founder of the Center for Marine Conservation, retired after 20 years of activism, he decided to become a consultant, not for NOAA, but for the Department of the Interior, believing the National Park Service is more committed to protecting the Blue Frontier than the nation's designated ocean agency. "I doubt NOAA is viable at this point," he told me more with a sense of resignation than anger. "NOAA's leadership believes they're there to develop marine resources and that kind of commercial thinking permeates the agency. They use the rhetoric of sustainable development but their biologists and scientists tell us that when they try to warn their superiors of problems, nothing happens."

And while NOAA officials like to claim that they must be doing something right because they are being attacked by both fishermen and environmentalists, it could just as easily be argued that neither group favors political opportunism in lieu of effective management. "NOAA's run by scientists lobbying for funding for their own institutions," says Zeke Grader of the Pacific Coast Federation of Fishermen's Associations, "so it's big on weather and hard science but not so good on resource protection."

I've heard similar comments from the head of a major oceanographic institute, politicians, congressional staffers, local and state coastal officials, ocean-industry entrepreneurs, and NOAA employees.

What the edge city of Silver Spring, Maryland, lacks in charm it makes up for in low-rent office space for second-tier government agencies like the $2.9 billion National Oceanographic and Atmospheric Administration. NOAA headquarters is a series of three glass, brick, and concrete office towers strung along the east-west highway (a fourth tower was evacuated several years back due to sick building syndrome).

Halfway up the block is a sculpture of a giant hand releasing four bronze seagulls—soaring off, possibly in search of bycatch to feed on. Inside this complex and throughout coastal America, I have had the opportunity to meet dedicated NOAA employees trying to do right by the Blue Frontier, but I have also found many of them frustrated and demoralized by the institution's misplaced priorities, misplaced plans, and often literally misplaced reports, schedules, and calendars. This institutional drift, along with its frequent internal restructurings, has led to a popular insider translation of NOAA: No Organization At All.

Without a strong advocate for the oceans, the Blue Frontier has tended to be defined as a marine treasure chest available to whoever has the political creativity to pry it open. For example, in 1998 Senator Ted Stevens of Alaska sponsored the American Fisheries Act. The bill was originally based on ideas put forward by shore-based Alaskan fishermen, Alaskan fish processors (mostly Japanese-owned), and Greenpeace, the environmental group. It was aimed at eliminating the giant Seattle-based factory trawlers that were competing with the Alaskans for Bering Sea pollack, which now makes up half the total U.S. catch by weight. The largest of the factory trawler companies, American Seafoods, was a subsidiary of Resources Group International (RGI), a Norwegian multinational that controls 10 percent of the global whitefish market. Other factory trawler operations included Arctic Storm and Trident Seafoods, which is partially owned by the grain giant ConAgra. In 1999 Trident bought out Tyson Foods seafood division, further consolidating the factory trawler industry.

When the American Fisheries Act was introduced, the largest capacity fishing boat in the world was tied up to a Seattle pier. RGI's *American Monarch* is capable of catching and processing about a million pounds of fish a day, using a net that could easily swallow the Statue of Liberty. This $65 million, 311-foot-long super trawler was denied permits to fish off Chile, Peru, and the Falkland Islands by governments fearful that it would quickly deplete their waters of fish and then move on. Because the *Monarch* was built in Norway (and U.S. law states that at least the hulls of fishing boats have to be domestically built) the *Monarch* was also excluded from the U.S. Bering Sea pollack fishery.

Meanwhile, the introduction of the American Fisheries Act in Congress was drawing lobbyists like dead fish draw cats. Along with lawyers and former congressmen, American Seafood hired Ted Stevens' brother-in-law, Anchorage attorney William Bittner. By the time the bill had gone through a series of closed-door meetings on Capitol Hill with the At-Sea Processors Association and other factory trawler lobbyists, it had been transformed from a bill to abolish factory trawlers into an industry subsidy. Restructured as a $97 million fishing boat buyback program, it allowed the industry to retire nine obsolete ships while consolidating its operations at sea. Shortly after the bill passed into law, Trevor McCabe, the Stevens staffer who oversaw the legislative deal making, quit the senator's office to become executive director of the At-Sea Processors Association with a considerable raise in pay.

While some advocates of the law argue that the American Fisheries Act helped reduce America's overcapitalized factory trawler fleet, it is worth noting that RGI and other global fishing companies also maintain pollack operations on the dangerously overfished Russian side of the Bering Sea, targeting the same fish as their U.S. boats. Recently, the Coast Guard, which protects the imaginary line in the water separating the U.S. and Russian pollock grounds, responded to a request that one of its rescue helicopters help transport an injured Russian crewwoman from the Vladivostok-based "Russian" fishing boat, *American Monarch*. The same giant factory ship that had been banned from Latin American and U.S. waters is now working the Russian side of the pollack line.

And then there's the greener side of offshore oil drilling. Room 366 in the Dirkson Senate Office Building is crowded with some 200 people. The high-ceiling, wood-paneled hearing room is rapidly filling with TV cameras, producers, print and radio reporters, and still photographers. The camera people are crowded below the dais at the front of the room where some 15 mostly Republican senators have gathered to discuss Senate Bill 25, the Murkowski-Landrieu Act, to reestablish full funding for the Land and Water Conservation Fund (LWCF). Created in 1965, LWCF set aside some $900 million a year from offshore drilling revenues for parks and wilderness. The original idea was to offset the negative environmental impacts from marine drilling with onshore conservation. Beginning in 1980, however, Congress started to hijack the fund to pay down the national debt and promote various Pentagon pet projects. Roughly $12 billion that was supposed to go to national parks and state recreation was never spent on its intended purpose.

But with record surpluses and the 2000 elections looming, there is a strong push in the House and Senate to pass some kind of environmental legislation that can be taken to the electorate. There are of course two competing versions of LWCF reauthorization—Democratic and Republican. The Republican version, which is the subject of this hearing, has its origin with oil industry executives who sit on the Outer Continental Shelf (OCS) policy advisory committee of the Mineral Management Service. As far back as 1993, in a report entitled "Moving Beyond Conflict to Consensus," they suggested that a portion of LWCF revenues be shared with coastal states where OCS drilling is taking place. This would provide an economic incentive for additional drilling and, in the ecologically correct language now used by even the most beach-fouling of industries, "support sustainable development of nonrenewable resources."

Personally, I'm impressed that there is so much public and press interest in a complex issue like land and water conservation funding. I figure maybe this is what happens when real money gets put on the table.

The chairperson of the Energy and Natural Resources Committee, Frank Murkowski of Alaska, like a florid Nero, is seated between decorative, oversized, bronze Roman torches. He makes it clear that with Senate Bill 25 "we're going to have a continuity of support for OCS."

Still, some of his western colleagues are distrustful of any legislation that threatens to expand public parks. "As a westerner where the federal government dominates by 63 percent, let me suggest, Mr. Chairman, we don't need anymore of the king's land," says Senator Larry Craig of Idaho.

There's a sudden stir in the room. "This is probably the most notice this bill will get in its life," the reporter seated next to me says, nodding toward the witness table. Suddenly, I understand why all the people and excitement. There, taking a seat, is Denver Bronco MVP Terrell Davis in a gray pinstripe suit and glasses. He's come to testify in favor of city parks that will receive some of the restored funding. Cameras flash, tape rolls. It is another celebrity day on Capitol Hill. I feel partly compensated for having missed Brooke Burns of *Baywatch Hawaii* when she gave her testimony against wasteful finning practices in the U.S. shark fishery.

Senator Ben Nighthorse Campbell of Colorado grins and says he'll want to mention one of his constituents later on. But Senator Evan Bayh of Indiana can't wait, using his designated time not to speak to the bill but to go on about what a great football player and human being Davis is. "And he's from Colorado," a slightly miffed Campbell notes. Other senators now begin to suck up to the football star, but Larry Craig is not to be deterred. "I grew up on a large ranch and didn't understand the importance of city parks, so I'm now sensitive to urban needs," he claims. "But I don't want the federal government owning one more acre in Idaho. I'm mainly concerned because federal lands become king's lands."

I wonder when the United States became a constitutional monarchy and how I missed that vote.

The first witness panel is led off by Victor Ashe, mayor of Knoxville, Tennessee, representing the U.S. Conference of Mayors. He tells the hearing of a pledge that was broken, that state and local LWCF funding has not been available to municipalities for too many years. The senators do not actually tap their fingers during his statement, but neither do they ask a lot of probing questions when he's finished. They know what will and won't make the evening news. As soon as the mayor's done, Senator Campbell gives a long introduction to the Broncos player, asking Terrell Davis's mom to stand up and take a bow. Davis, testifying on behalf of Pop Warner scholars and the Sporting Goods Manufacturers Association, explains how LWCF funding will provide needed urban parks and youth sports playing fields, like the one he used when growing up. When he concludes his remarks a dozen still and TV cameras flash and roll. The next speaker, Bernadette Castro from the New York State Office of Parks, offers Davis a Mont Blanc pen, a gift from her children, for all the good works he's done. I wonder what her kids think about her giving away their gift. In her statement she mentions the importance of coastal conservation and how New York's Jones Beach sees more visitors every year than Yellowstone and Yosemite National Parks combined. "A promise was made in 1965," she points out, "and that promise was broken."

The secretary-general of U.S. Soccer and the president of the city of Dallas Parks and Recreation also speak in favor of renewed funding.

During the follow-up questioning, Murkowski asks Terrell Davis about quarterback John Elway's retirement. Senator Conrad Burns of Montana tells a joke about a young football player with a glass eye. His coach, after a risky play, asks the player, "What if he'd poked you in the other eye? What would you do then?"

"Then I'd become a referee," the player replies.

I'm beginning to think term limits might not be such a bad idea.

Suddenly, Senator Pete Domenici of New Mexico decides to score some political points. "I'm not impressed with buying more land in the inner city," he says. "You need adults to work with kids, volunteers, and trainers, and that's the important thing." He asks Terrell Davis if he doesn't agree, but Davis isn't going to let himself be sacked so easily.

"I was lucky to grow up between two schools and a park, and we organized our own games and we kids played till it was dark so we were too tired to do anything but go home then. . . . We did it ourselves because the park was there."

The first panel ends and the cameras are in Davis's face, and even though I'm a 49ers fan and never much liked the Broncos, I think I'm now a Terrell Davis fan as well. Some people linger, asking for his autograph as two-thirds of the audience and three-quarters of the media file out of the hearing room.

Murkowski convenes the second panel, led off by Chuck Cushman, a stalwart of the anti-environmental Wise Use movement, known to his followers as "Rent-a-riot." When Chuck begins testifying about cultural genocide, comparing the impact of national parks on rural Americans to Serbian ethnic cleansing and atrocities in Kosovo, I decide it's time for me to leave, too.

It's Clean Water Day in San Diego and members of Surfrider, the eco-surfer group, are getting ready to paddle out and around the quarter-mile-long Ocean Beach pier. I'm torn between my camera bag and my bodyboard, between taking pictures and joining the sea-besotted throng on the familiar white sand of my old but little changed neighborhood. Although it's 7:30 in the morning, the beach at the foot of Newport, Ocean Beach's main palm-lined business street, is already crowded with some 500 surfers along with TV trucks, their microwave masts raised for live morning-news feeds, a tent set up by 91X rock radio, several straw palapas peopled by Surfrider, and traffic cops on mountain bikes and in patrol cars.

Mixed in among the crowd I find my old friend Jeff Stone and his young sons, Cody and C. J., wetsuited and ready to hit the water; and Donna Frye and Brian Bilbray, two folks who get along about as well as your average mongoose and cobra. Donna is a former antiwar activist, union organizer, feminist, and founder of STOP—Surfers Tired of Pollution. She is also the wife and partner of surfing legend and custom-board shaper Skip Frye. Brian is a Republican congressman, surfer, father, and ex-mayor of the biker border town of Imperial Beach, where he once created an international incident by attempting to block the flow of sewage from the Tijuana River with a bulldozer.

Donna is lanky and tanned, with aquiline features, long auburn hair, and a slightly goofy smile that reminds me of Cecil, the seasick sea serpent, one of my favorite TV puppets when I was young. Surfers, male and female, keep coming up to say "hi" or hug her, including Glenn Hening, one of the founders of Surfrider. Skip, his wetsuit top unzipped and hanging at his waist, longboard under one arm, comes over to place a green ti lei around his wife's neck, which just lights her up.

Representative Brian Bilbray is fortyish, boyishly good looking with fine blond hair, a trim surfer's body, and easy, energetic smile. He's being interviewed by a reporter-cameraman from Channel 10, while his 14-year-old son fidgets beside him. They're both ready to hit the water, as are several of his staff aides. He's telling the newsman about the BEACH bill, the Beaches Environmental Assessment, Cleanup, and Health Act, that he has introduced in the House. It requires testing the waters at all public beaches in the United States for bacteria, viruses, and other pollution risk factors. It's a bill endorsed by Surfrider, the American Oceans Campaign, Center for Marine Conservation, and other blue groups. It also reflects a sea change in Bilbray's political positioning since he was first elected in 1994.

"In 1994 I got very vocally active because of Bilbray and Newt Gingrich and their attempt to gut the Clean Water Act," Donna tells me. "I was dealing with a lot of sick surfers at the time, Skip being one of them, and it didn't make sense. These are healthy, athletic people who were getting sick from pollution. And what was doubly insulting is Bilbray ran against [Democrat] Lynn Schenk and said, 'Vote for me 'cause Schenk don't surf,' and the implication was that she didn't care about the ocean and so he used the surf community to promote his own political agenda."

The 1995 Clean Water Act "reforms" proposed by the House Republicans (which was labeled the "Dirty Water Act" by its critics) were written by lawyers for the oil and chemical industries, with help from the U.S. Chamber of Commerce. The proposed reforms zeroed out funding for the EPA water quality monitoring program, redefined wetlands to eliminate 80 percent of them from legal protection, exempted the navy's irradiated discharges from the legal definition of nuclear waste, suspended programs to control sewage and storm drain overflows, and created a waiver for secondary sewage treatment if waste water was discharged directly into the deep ocean as it still is in San Diego.

"I don't think it was good for the whole country but parts of the bill were essential for San Diego, to save a billion dollars for my constituency," Bilbray tells me. "It was the only vehicle we had at the time to do that [stop the EPA from requiring San Diego to build a secondary treatment plant]."

As a result of his vote, Bilbray came under fierce attack from groups like STOP, which operated out of Harry's Surf Shop in Pacific Beach. STOP printed bumper stickers reading, "Another Surfer Against Bilbray & for Clean Waters." They began showing up on cars at all the area surf spots. Bilbray's pals responded with stickers reading "STOD—Surfers Tired of

Donna—Truth Was Her First Victim." The PBS documentary *Fender Philosophers* gave their war of words some play as did *Surfer* magazine (which had them debate) and the San Diego media.

My old roommate Charlie Landon, a surfer and cameraman for the local CBS affiliate, went on assignment with a reporter to interview Bilbray about his environmental stance. After the interview Charlie asked when was the last time he'd been out in the water.

"Not for a while," Bilbray admitted. "Last time I was out I got this ear infection." Charlie and his reporter just looked at each other.

As the Republican environmental agenda came under fire, Jim Saxton, Sherry Boehlert, and other party moderates drafted more conservation-oriented Clean Water Act language, but their amendments were defeated in a House floor vote. Bilbray wrote a letter to his constituents pointing out that he had voted for these greener measures. But according to the *Congressional Record* for May 10, 1995, Bilbray voted with the majority against Saxton and Boehlert's amendments, only changing his vote after it was clear they would not pass. However, as Bilbray's victory margins began to shrink in subsequent elections, his rhetoric and voting record also began to shift.

"Bilbray is trying to get on the right side with environmentalists because Surfrider and Donna beat him up on the dirty water vote," explains Barbara Jeanne Polo, executive director of the American Oceans Campaign. "The 1998 election was a close call for him. So he told us he wants to move forward on the BEACH bill and run on it in the next election. He told us this is the way to go and whatever input we want on it is fine."

"You can't not care about the ocean, growing up in Imperial Beach," Bilbray tells me. "My son, he's a second-generation sewage kid. But you know, Donna wouldn't support me if hell froze over. She doesn't know anything about me. I don't know what her involvement is, but I'm the guy to go to if you want to do something about these problems. How many mayors or congressmen have had their Miranda rights read to them?" (That happened following a second bulldozer incident in 1990.)

Bilbray, Skip Frye, Jeff, Cody, C. J., and about 495 others paddle out into the water, forming a sinuous broken line like some colonial animal swimming along the pier and around its wide T-shaped terminus. On the pier itself at least 100 more people, mostly Vietnamese and Latinos, are fishing for mackerel, bass, and queenfish. Their rods and throw nets almost reach to where the rafts of surfers are paddling by in a 2-foot swell. Few comment on the passing phenomena. These people are here to fish. A Vietnamese lady brings up three small perchlike queenfish on her three-hooked line. Her husband puts them in their iced bag with a dozen others— supper for tonight. Gulls and pelicans perch on the scarred wooden rails, waiting expectantly for a handout or fumbled fish. I look down along Sunset Cliffs to the brown clapboard house where I used to live. The sky is clouded over, the air fresh, tasting of salt and iodine. This is the Blue Frontier at its best, a large, diverse bunch of people taking pleasure and paying homage to what the god Neptune has to offer.

I talk with a dozen stoked surfers back on the beach. They've hooted and splashed and gotten wet in a good cause and the city council has even passed a resolution declaring this Surfrider Clean Water Day.

Unfortunately, six months later Ocean Beach will be hit with a 36-million-gallon sewage spill, the worst in the city's history, even as the city is suing the Environmental Protection Agency (EPA), claiming the feds don't have the right to make San Diego reduce the amount of untreated sewage it pumps into the ocean.

On the way out of the parking lot I stop and talk to one more surfer. Tom Sekreta works for a sporting goods company, is one of 30,000 dues-paying members of Surfrider, and has lived in Ocean Beach for 26 years, surfing for 25 of them. He has curly white hair, a friendly grin, and a large beer gut just to prove that not every surfer is studly. I ask him what he thinks of Brian Bilbray.

"When I look at Bilbray and consider he won his last election by less than 2 percent, I figure he has to listen to the RNC [Republican National Committee] and go where the money tells him to. He's got a thin veneer of doing this stuff but when you scratch it, you realize he could do more but instead has to toe the party line. It really all comes down to the need for campaign finance reform, I think. Otherwise it's just the money and the big polluters in charge."

But there is something more going on here, I suspect. In California people have a sense of entitlement when it comes to the ocean. Unlike New England, where people think the water belongs to the fishermen and the beaches to the townships, or Louisiana, where they know it all belongs to big oil, Californians believe the ocean is their birthright, or becomes so on acquiring residency. As a result, they regularly put pressure on elected officials like Bilbray to do right by the Blue Frontier. (Bilbray will still lose his seat in 2000.) They have also created and continue to support a model resource agency, the California Coastal Commission.

The world's first coastal management agency was the San Francisco Bay Conservation and Development Commission (BCDC), founded in 1965. Not surprisingly, it came about as the result of a Corps of Engineers plan to fill in San Francisco Bay. Even accounting for the can-do spirit of the times, it's hard to look at the 1959 corps map of the proposed landfill and not shudder. Richardson Bay in Sausalito, where I lived for six years, a sparkling arm of the larger bay blessed with houseboats, sailboats, egrets, great blue herons, tidal marshes, and occasional barking sea lions chasing herring, would have become an industrial flatland.

By the Beat era of 1959 San Francisco Bay was already a third smaller than during the gold rush more than 100 years earlier. With much of the water less than 18 feet deep, the bay was too easy to fill. The lower bay was diked off for salt ponds. Flying into San Francisco today you can still see the red, orange, brown, and purple colors that different algaes give to the big evaporation ponds. The north bay was reclaimed for agriculture and later by duck hunting clubs. Much of what is today downtown San Francisco was also built out from the shore. During the gold rush, sailors jumped ship to mine the foothills of the Sierras. Their abandoned sailing vessels were converted to jails, hotels, and brothels. Eventually, the ships and their piers rotted and sank into the bay mud, where they were covered over with rocks and dirt. Even though the bay's waters held high-value fisheries (author Jack London was both an oyster pirate and fish patrol agent on the bay), the infill continued well into the twentieth century. But the 1959 Corps of Engineers plan to fill 60 percent of what remained, a plan that would have transformed the bay into a wide spot on the Sacramento River, marked a turning point for how citizens of the region viewed their world-famous estuary.

Three women from Berkeley, including Kathryn Kerr, wife of the university president, formed a group called Save San Francisco Bay that quickly grew in strength, halting the corps' plans and getting the state to establish BCDC. By 1969 the state had empowered the new commission to regulate development on and around the bay and its 1000 miles of shoreline. Today, along with reclaiming historic wetlands and preventing new fill, BCDC is involved in opening up waterfront parks and trails, working on issues of waterborne recreation and commerce, and beginning to develop plans for improved and expanded ferry service. "Our weakness is our jurisdiction only extends 100 feet inland and we're not authorized to deal with issues like nonpoint pollution or water allocations," says BCDC executive director Will Travis. "Still, we became a model for the California Coastal Commission and similar groups in Oregon, Cape Cod, Japan, New Zealand, and elsewhere."

Along with the Corps of Engineers scheme to fill in San Francisco Bay, the 1960s saw plans to expand California's famous Pacific Coast Highway into a multilane freeway, build hundreds of new homes on what is now Point Reyes National Seashore, construct Miami Beach–style high-rises along the state's central and southern coast, drill for oil off Monterey and Big Sur, and install a nuclear power plant on scenic Bodega Bay headlands north of San Francisco.

By 1971 Peter Douglas, a legislative assistant in California's capital, Sacramento, had written up a bill to counter these threats by establishing a statewide coastal commission based on the BCDC model. Its main purpose would be to assure public access to and scenic protection of the 1100-mile-long California coast. The bill had the support of a broad coalition including the League of Women Voters, Longshoremen's Union, and Sierra Club.

"My boss got it through the assembly," Douglas recalls, "but the day the senate was to consider it, our key senate vote didn't show. I ran into a lobbyist in the hallway who said the state senator was flying to his ranch to take delivery of a race horse he'd gotten from another lobbyist who worked both for the Racing Association and several oil companies. I called the press and they sent TV cameras to his ranch. This horse truck approached and when the driver saw the cameras he did a 180-degree turn."

Frustrated by the corrupt legislative process, Douglas and his friends decided to go the initiative route, putting coastal protection on the 1972 ballot as Proposition 20. Even though the Homebuilders Association and other developers outspent Prop 20 supporters about 100 to 1, they were unable to counter the state's nascent environmental movement or the public's memories of the Santa Barbara oil spill three years earlier.

"We won not with money but press coverage of coastal damage that was taking place at the time," recalls Congressman Sam Farr of Monterey. "We only had a staff of three—Peter, Bill Press, who's now on CNN's *Crossfire,* and me. I led a bike group from San Francisco to San Diego, and all along the ride we'd stop and explain to people why we needed Prop 20."

On November 8, 1972, the coastal initiative won with 54.5 percent of the vote. To prevent the commission from becoming the tool of any single politician or agency, the initiative's authors required voting membership be divided up, with four commissioners appointed by the speaker of the house, four by the senate rules committee, and four by the governor. The commission in turn required every coastal county in California to develop and periodically upgrade a plan to guarantee local protection and access to the coast.

Today, the California Coastal Commission takes up the nineteenth and twentieth floors of a downtown San Francisco high-rise. It is here that I meet with Peter Douglas, the now bald, gray-bearded, and avuncular executive director of the commission, a position he has held since 1985.

"There's no doubt our state has the most accessible coastline in the country, because of the public's activist and outspoken concern in terms of protection," he claims. "Without strong coastal commissions and local plans in states like Texas, Maryland, and Florida, you see these coastal seawalls emerge, made up of endless miles of waterfront high-rise hotels and condominiums.

"When Vandenberg Air Force Base converted to a space launch center and wanted to bring in a water pipe, we said no," he tells me, citing one example of the commission's work. "We knew that extra water would have a growth-producing impact because it's so dry along the coast there. So, we required the base to establish a water conservation program so they wouldn't have to bring in new water, and we required they open up several miles of beach to the public [it's a good surf spot] and that they not launch missiles over the Channel Islands during seal pupping season when it might effect the seals, and they agreed to all that."

Later in the 1990s, when the army decommissioned Monterey Bay's Fort Ord, the California Coastal Commission made sure that its billion-dollar beach west of Highway 1 went into the state park system rather than to private developers.

Under rules of the federal Coastal Zone Management Act, the commission has also fought offshore oil, prevented the navy from scuttling old nuclear submarines off the coast, and blocked the EPA from giving permits to toxic incinerator ships. "Provisions of the act that allow the states to participate in federal decisions impacting their coasts has allowed California to reach out into the EEZ," Douglas says with a somewhat acquisitive grin.

But it was forces closer to home that almost scuttled the commission back in 1996, when Republican governor (and erstwhile presidential candidate) Pete Wilson and the Republican-dominated state assembly and senate packed the commission with real-estate developers and property rights activists, before asking Douglas to hand them his letter of resignation.

"What happened is they came to me and asked me to recommend building 900 units of housing in Bolsa Chica [an Orange County wetland being developed by the Koll Corporation], and when my staff said they wouldn't recommend building homes in a wetland, Doug Wheeler, Pete Wilson's secretary of resources, became furious. They also wanted the commission to allow Southern California Edison to escape mitigation requirements they'd already agreed to when they built the San Onofre nuclear power plant."

Douglas's recollection of what set off his attempted ouster was confirmed by reporters from the *Los Angeles Times* who interviewed Wheeler and other principals. Douglas asked for a few weeks to consider his resignation, then asked for a public hearing. By the time the Coastal Commission met in Huntington Beach, there had been an outpouring of public support for the executive director and his staff. Letters supporting Douglas came from public officials and county governments up and down the coast. Thousands of letters and phone calls of protest flooded Sacramento, demanding the Coastal Commission get back to the job of protecting the coast. Angry editorials appeared in every major newspaper in the state. "An Endangered Coast," warned the *San Francisco Chronicle*. "Do you want the coast to be spoiled? Apparently Pete Wilson's administration does," harked the *San Jose Mercury*. The *Sacramento Bee* labeled Douglas "the coast's best friend," while the *Los Angeles Times* warned Wilson that "California's irreplaceable coast is not a political pawn." Editorial cartoonists tended to portray the state's political leadership as either pirates or sharks.

The *Los Angeles Times* also reported, "Fear of the potential political fallout for Republican candidates in the November election sparked at least three GOP lawmakers from coastal districts to oppose the move against Douglas."

Hundreds of Douglas's supporters (including Alexandra Paul of *Baywatch*) packed the Huntington Beach meeting on a warm Friday evening in July, calling the prodevelopment commissioners "cowards" and big-money "shills" until Douglas stood up and pleaded for calm.

Amidst boos and catcalls, the commission voted to postpone its decision on firing him. "You haven't got the guts to do this in front of everybody here," liberal holdover commissioner Sara Wan of Malibu yelled at commission chairman Louis Calcagno, a Pete Wilson appointee. Calcagno responded by suggesting that a simple management personnel issue had been "turned into a circus by some in the media and by vocal special interest groups."

In the wake of that meeting, Sara Wan helped organize Vote the Coast, a political action coalition that targeted 10 races in that fall's election, supporting winners in eight of them. This effort proved pivotal in putting Democrats back in control of the state assembly. Coastal protection became an even larger issue in the 1998 election, when Vote the Coast candidates won all their races and Democrat Gray Davis made coastal protection a top issue in his successful run for the governorship.

"California is a place where you can still get elected running against offshore oil and for protection of the coast," explains Congressman Sam Farr, who has done it himself.

Today, Sara Wan is chair of the Coastal Commission, and the Bolsa Chica wetlands, through a cooperative agreement between the developer, local environmental activists, and a statewide land trust, has been sold to the state as a coastal wildlife reserve.

"Wheeler and Wilson tried to drive us off a cliff, and it backfired and now we probably have a stronger commission than we've had since the 1970s," says an unabashedly pleased Peter Douglas.

On March 7, 2000, amidst a swelter of 21 ballot propositions, the people of California passed two: the Parks, Water, and Coastal Protection Act, and the Safe Water Act, which committed more than $4 billion to, among other things, protecting watersheds to reduce nonpoint pollution, improving the quality of coastal bays and beaches, and assuring clean water for drinking, recreation, and wildlife, including ocean fish and marine mammals.

This suggests that despite the growing crisis of our living seas there still are, along with unresponsive bureaucracies and unending red tape, signs and centers of hope where people can take on a greater stewardship responsibility for their coasts and the waters that lie beyond.

If part of America's glory can be found in its still vast wilderness ranges and national parks from Alaska to Yellowstone to the Everglades, what about the possibility of moving offshore to create a similar system of large reserves and underwater wilderness parks? Luckily that process is also underway.

Chapter Eleven

Sanctuaries in the Sea

There is, one knows not what sweet mystery
about this sea, whose gently awful stirrings
seem to speak of some hidden soul beneath

—HERMAN MELVILLE

Where great whales come sailing by,
Sail and sail, with unshut eye,

—MATTHEW ARNOLD

In its early years Yellowstone National Park counted among its more popular activities trophy hunting for elk and bison. Today, America's national marine sanctuaries are in their early years.

Florida Marine Patrol (FMP) officer Greg Stanley pulls up to the dock in a sleek Olympia. When the FMP seized this former drug-running muscle boat, it was registered as a 13-foot canoe. The 31-footer with twin 225-horsepower outboards can actually do better than 55 miles per hour on the water, faster than almost anything out there except newer drug runners and Mark V navy SEAL boats.

Tall, with a blond buzz cut, angular clean-cut face, dark glasses, shorts, and a Kevlar vest under his short-sleeved uniform shirt, Greg wears dual shoulder patches—for the FMP and the Florida Keys National Marine Sanctuary, for which he's been cross-deputized.

Today is the opening of the sanctuary's two-day lobster miniseason, which precedes the sanctuary's commercial lobster season, and draws tens of thousands of lobster-hungry hobbyists from throughout Florida and the South. Every boat on the water is permitted six lobsters per person. Unlike California, where you have to work for your lobsters by catching them by hand, a skill that has left me humiliated by fast-moving crustaceans on more than one occasion, here you can use tickle sticks, nets, nooses, everything but dynamite to make sure you get the limit. It reminds me of the meat fishery for hatchery salmon in Homer, Alaska, except no one is restocking these animals.

Greg and I leave the Boca Chica Basin north of Key West, jigging and jagging through the mangroves, the wind and engines screaming as Greg leans the go-fast boat over in tight g-pulling turns. We head out into the 9.5-mile SPA (special protected area) and find little activity. Fishers and pirates are learning to stay out of the no-take zones, which cover less than 3 percent of the sanctuary. In 1990, when the Florida Keys National Marine Sanctuary was established, biologists suggested 20 percent be declared a no-take zone to protect the living resources of the reef. There was a tremendous backlash from charter boat operators, treasure hunters, salvagers, and tropical fish and live coral collectors. Sanctuary manager Billy Causey was hung in effigy at a series of angry rallies backed by the late treasure hunter Mel Fisher and his attorney. The feds retreated. Ten years later, in March 2000, the U.S. Coral Reef Task Force, created by presidential order and including governors of states and territories containing coral reefs, issued a new report suggesting that to save America's reefs, at least 20 percent be set aside as no-take zones. With the continued biological decline of the Florida Keys, a solid majority of the area's residents now support increased protection.

Greg tells me that tropical fish poaching is a problem in the SPA because the aquarium trade pays such high prices for the jewel-like little fish. On this patrol, however, he's looking for undersized or over-the-limit "bugs" (lobsters). There's a $210 fine for the first five undersized lobsters, an amount that jumps to a $315 fine for six to a dozen. "The guys I really want are the big poachers with 125 lobster tails on board but they all came out last week. They know today will be heavily patrolled."

His first inspection is of a little green bumper boat that looks like a refugee from a carnival ride. A father and his young daughter are on board. The only problem Greg sees is one of safety. The man has two adult lifejackets, neither of which will fit his 75-pound little girl. "Square it away," Greg advises, declining to issue a ticket.

We next stop by the *Filet & Release,* an open dive boat with two guys in the water and a woman waiting on board with fresh beer and lobsters on ice, ready for a good day. We next check out a family of eight, with three kids and 14 lobsters, four navy guys and a gal with 16 lobsters, six people with a cooler full of lobsters and three speared hogfish, all easy kills they tell us.

Greg next stops a poorly maintained 25-foot Boston whaler with a Bimini shade top off of Snipes Point. On board are two boys, a teenage girl in too-tight jeans and a halter top, an older woman, and an unshaven gray-haired man with a large beer gut, orange T-shirt, and greasy jeans. There's also two divers in the water. Greg begins spotting undersized lobsters and throwing them onto his boat.

"Those shorts, daddy?" the nearest scuba diver calls from the water. He's also gray-eyed and gray-haired, a construction worker from Key West. Greg calls him over by our boat and asks to see his capture bag. He swims up alongside. The bottom of the net bag is open. Only one lobster is left clinging to its side.

"How'd your bag get open?" Greg asks.

"I don't know."

Greg takes his measuring gauge and finds the lobster stuck to the side is legal, if just barely.

"You have to measure from between the eyes?" the diver asks unconvincingly.

Greg begins writing them up for five undersized lobsters.

Then Dad tosses his cigarette in the water.

Greg just stares for a second or two. "Don't put that in water. That's littering, sir. Did it have a filter?" Dad shakes his head vaguely. "A filter will never dissolve," Greg explains before taking Polaroids of the confiscated bugs, writing our GPS position on the back of the photos, and then tossing the lobsters into the water where I watch them skitter away into the eel grass. Maybe they heard what he had to say about the filter.

"That guy had a bagful he dumped. Nothing I could do about it," Greg says as we speed off.

The next boat he stops is a cabin cruiser with four adults, four kids, and an unbelievable amount of toys—coolers and a barbecue, snorkel and dive gear with eight air tanks, an expensive underwater scooter, two rafts, and six legal and five undersized lobsters. Greg has to shake several loose from his glove, dropping them hard onto his boat.

The big boat's owner, a musician, is about 6 feet tall, fleshy, with a T-shirt reading "The older I get the better I was." He seems offended to be getting ticketed. We drop the lobsters back into the water. I watch one fall then right itself and take off like a rocket.

"I shouldn't treat the lobsters so rough," Greg worries as we pull away. "It doesn't hurt them, they're tough critters, but it makes it look like I just want to give tickets and don't care about the environment, which I do."

We stop to help a couple of men in the no-take zone, trying to figure out their charts. We then check out some other boats, including a classic 1960s cabin cruiser with two geezer gay guys and their Jack Terrier. The 80-year-old tells us he's not interested in lobster but might go for a swim later. A few dolphins glide past us in the translucent aquamarine water. Greg pulls up next to a big sportfishing boat named *E-Fish-N-Sea*. A second boat and raft are tied off behind it. There are nine on board the main craft, including an infant.

"You count for six [lobsters]," the baby's aunt coos, holding her up for our inspection. There are 37 lobsters in one boat, 17 in the other, and they want to know if they can catch more tomorrow since they're staying onboard overnight. "Only if you leave these ones on land," Greg explains the rules.

Looking across the crowded coral flats, it's easy to believe that there are now 17 million recreational boats plying America's waters (including 2 million sailboats). Half the fleet seems to be in the Keys today hunting lobster. Nearby a motorboat is towing a pontoon boat towing a Jet Ski.

Greg talks to three people with 19 lobsters instead of 18 and judges the extra lobster an honest mistake. "You should count and recount again," he reminds them before tossing one lucky lobster back into the sea.

He pulls up to a pair of tanned, short-haired, 16-year-old boys having engine problems. They also have 21 lobsters and some undersized fish on the bottom of their blue skiff. He takes all but 12 lobsters and tells them he's writing them juvenile citations. "You'll end up doing community work hours. You won't get a criminal record but in a few years you could end up with a record over something as stupid as lobster if you keep doing this," Greg warns.

One of the boys calls his dad on a cell phone, tells him they've been stopped and are being cited. The boy listens, then says, "Some are on his boat. We still have 12."

"Most Keys residents don't like miniseason. It teaches people how to rip off and abuse the reef," says Dave McDaniel another sanctuary patrol officer I take a ride with. Opposition to the miniseason ranges from ecoactivist groups like Reef Relief to the Sanctuary Advisory Council and the Florida Keys Tourist Development Board. Still, it is up to the State Fisheries Commission to determine whether the annual meat fest will continue, be folded into the commercial season, or be phased out.

Unlike most national parks, America's twelve national marine sanctuaries share jurisdictions with states where their waters overlap. Two-thirds of the 220-mile-long Keys Sanctuary is in state waters. The sanctuaries also allow a range of commercial activities within their boundaries, including commercial fishing, cable laying, and oil transport.

The National Marine Sanctuary Program was created in 1972, 100 years after Yellowstone was dedicated by Congress as the first national park. Unlike the establishment of Yellowstone, which involved a high-profile campaign to open up the wonders of the interior West to railroad tourism, the sanctuaries enabling legislation was passed as a rider to an ocean-dumping bill, one of a number of environmental acts passed by Congress in the early 1970s. According to the act's original language, sanctuaries are supposed to protect the "conservation, recreational, ecological, historical, research, educational, or aesthetic qualities" of America's Blue Frontier, allow for multiple use of the marine environment, and protect the long-term integrity of its natural resources. Beyond excluding oil and gas drilling, mineral mining, and ocean dumping, how marine sanctuaries might achieve all these aims is an open question.

The push to create marine sanctuaries was in large measure a response to public anger and frustration over the disastrous Santa Barbara oil spill of 1969 and reported dumping of military nerve gas and nuclear waste off the East Coast. The oil and gas industry opposed the establishment of the sanctuary program as did the Departments of Defense and Commerce (even though NOAA would be given charge of them). While appearing to stand firm against the pressure, Congress nonetheless failed to provide the new program with any funding during its first seven years of existence.

In 1975 the first marine sanctuary was established 16 miles off Cape Hatteras, North Carolina. It was a 1-square-mile box reaching down 230 feet, to where a Duke University research vessel had discovered the overturned wreck of the USS *Monitor*. The ironclad warship had been built in the Brooklyn Navy Yard, then towed to Hampton Roads, Virginia. It arrived on March 6, 1862, in time to stop the CSS *Virginia* from finishing off the wooden ships of the port's federal blockading force. After an historic, deafening, but indecisive four-hour battle on March 9, both steam-powered ironclads withdrew from the scene. The *Virginia* was scuttled when the Confederates abandoned Norfolk. Months later, on the last day of the year, the *Monitor* sank in a storm while under tow to Charleston. Sixteen of its crew drowned. It apparently was involved in one more engagement; it was depth-charged during World War II by a surface ship mistaking its sonar signal for that of a Nazi submarine. Researchers and scientists from NOAA and the navy are working on the coral encrusted, rapidly decaying shipwreck, hoping to recover its turret for display at the Mariner's Museum in the former Confederate port town of Newport News, Virginia.

At the end of 1975 a second national marine sanctuary was established to protect a fragile coral reef habitat at Key Largo, Florida, a site that had been under a presidential protection order since 1960. Both of these original sanctuaries were financed by monies diverted from other NOAA programs.

Beginning in 1977, President Jimmy Carter accelerated the designation process, leading to the establishment of four new sanctuaries off Florida, Georgia, and California, including the 1658-square-mile Channel Islands Sanctuary, whose waters are frequented by 22 species of whales and dolphins, at least 20 species of shark, and some 150,000 sea lions.

During the eight years of the Reagan administration only one mini-sanctuary was established, at the 162-acre Fagatele Bay in American Samoa. Political appointees within NOAA who oversaw the sanctuary program made it clear that they were not going to impede the plans of Secretaries of the Interior James Watt and Donald Hodel to open up a billion acres of the Outer Continental Shelf to oil and gas development. These plans included expanding the number of drilling platforms off the coast of California from 20 to 1100 and opening up New England's historic fishing grounds and Florida's coral reefs to drilling for the first time. What Congress, the administration, and NOAA's political appointees failed to count on was an explosion of grassroots outrage over the proposed leases. In central California, antidrilling groups like Save Our Shores linked up with commercial fishermen, local governments, and the tourist industry to lobby for permanent protection of the coastline.

"We'd really taken our oceans for granted till Watt and Hodel made people sound the alarm and got us thinking about what we could do," recalls Representative Sam Farr, who was a Monterey county supervisor at the time. "We'd never even thought they'd drill for oil off northern California and then MMS put out the call and 27 oil companies said they were interested in offshore Santa Cruz. So we organized a rabbit out of a hat. The federal government had created this opportunity for regions to create marine sanctuaries. So, we proposed Monterey and we had the canyon as this unique thing, but couldn't get the maps of it, 'cause the navy still had them [classified]. So what we did then is change the business versus environment argument into a business versus business deal.

"We talked to agriculture and looked at the Mediterranean climate that makes this area ideal for growing strawberries and all these other crops, and the farmers said we don't want hydrocarbons that could change the air quality, polluting our air from drilling operations. And we went to the tourist industry and that was an easy sell, because people came to the coast for the views and the water and we looked at our county and said, 'We have $2.4 billion a year in agriculture and $1.8 billion in the tourist economy, and this is too high a risk for us, as against so many barrels of oil they might drill. So we took out ads in the *New York Times* and got a drilling moratorium through Congress, and I remember thinking we're the mouse that roared."

"I'd never seen such a broad coalition form so quickly in my life. It was real groundswell," agrees Leon Penetta.

Penetta, who was Monterey's congressman at the time (and became White House chief of staff during the Clinton administration), added a rider to the Hurricane Andrew Relief Act of 1992 to create a new sanctuary of uncertain size. The sanctuary's proponents drew up maps for a small-, medium-, or large-sized reserve. The largest would cover more than 350 miles of coastline and extend as far as 53 miles out to sea. At more than 5300 square miles, it would be the biggest marine reserve in the world after Australia's Great Barrier Reef.

"Leon said let's compromise on the middle one, but the people just went nuts at all these public hearings and insisted on the biggest deal," Sam Farr recalls. "Luckily '92 was an election year, and we reminded George Bush that in '88 he'd run a TV ad in California showing the Big Sur coastline, saying he was going to be the environmental president, so he went for the biggest boundary." (Bush would still lose California to Clinton in the November elections.)

On September 19–20, 1992, the Monterey Bay National Marine Sanctuary was dedicated. Monterey's Shoreline Plaza was crowded with visitors checking out the more than 50 booths at the sanctuary celebration. On display were baleen from humpback whales, decompression tanks for divers with the bends, fiberglass kayaks, search-and-rescue arrays, power boat cutaways, hands-on tide pool animals from the Monterey Bay Aquarium, and a tethered observation balloon from NOAA. The Surfrider Foundation set up a tubular 7-foot wave made of painted plywood with a surfboard attached to the inside face. Penetta and anti–oil drilling activist Dan Haifley rode tandem on the board, pretending to shoot the tube as they mugged for the cameras. Beyond a row of environmental booths with displays of wildlife, seashells, and marine litter was an exhibit of schoolchildren's artwork—pictures of sand castles, dolphins in hula skirts, and breaking surf represented by blue waterpaint and glued-on white cotton balls. "Did you pet the stuffed otter?" a young girl asked her brother. "It wasn't stuffed, it was freeze-dried," he replied knowingly. There were gulls cawing overhead and other kids barking back at the seals in the blue-green water below the seawall. Four

teenage lifeguards from Santa Cruz paddled into view. They had just completed a 26-mile journey by longboard across the frigid waters of the bay,
encountering wild dolphins in the fog along the way.

At 10:30 A.M. a parade of boats arrived, led by the 145-foot
Californian, a replica of an 1849 revenue cutter. Behind the tall ship were
modern-day Coast Guard cutters and patrol boats, an oceanographic
research vessel, a fishing trawler flying Old Glory, sailboats, kayaks, and
Zodiacs. The Monterey Bay Symphony struck up Copeland's "Fanfare for
the Common Man" as the *Californian* fired a loud, smoky volley from its
guns. The gunfire did not seem to deter dozens of curious sea lions that
were leaping and nosing their way among the somewhat nervous kayakers.
The woman next to me turned to her husband. "How'd they get them to do
that?" she wondered.

Years have passed, but my wonder remains. It's an overcast winter day
on the waters of Monterey Bay as I gear up with NOAA Corps Lieutenant
Mark Pickett; Ed Cooper, the diver representative to the Sanctuary
Advisory Committee; and Carrie Wilson from California Fish and Game.
We're all in full wetsuits with hoods and booties, except for the boat's captain in the partially enclosed cabin who's wearing jeans and a hooded sweatshirt and looking distracted as he backs off the anchor chain. Suddenly,
there's a splash and we all turn to see a large smooth footprint on the surface. It's right next to our 28-foot boat and about the same size.

"Did you see it?" I ask. "No." "No." "How about you?" come the replies.
None of the five of us has seen whatever it is that just left this watery mark,
whose rippling edge is now lapping up against our hull. We all look toward
the kelp bed to our seaward side, no one making any motion toward the dive
platform. Nor is anyone saying what we're all thinking. Monterey is the
southern point of California's infamous Red Triangle, which extends north
past Stinson Beach and out to the Farallones Marine Sanctuary. It's the
world center for human–white shark "encounters," and California's white
shark population is, according to experts, "robust," which is supposed to be
a good sign, the presence of top predators being an indicator of a healthy
ecosystem. Just as our silence begins to get uncomfortable, a juvenile gray
whale surfaces snout first on the other side of the kelp. It gives a loud
steamy blow of air. We grin and point, checking out the huge young animal,
then checking out our masks and regulators and stepping off the dive platform into the bracing 55-degree water, flipping over and heading down
along the anchor chain.

Below the surface a large shoal of bluefish hangs suspended amidst
yellow-green stalks of 65-foot-tall giant kelp rising up from their holdfasts
on the bottom. Pinnacles of rock also rise from a bottom littered with orange

starfish, faster-moving sunflower stars, and purple-black sea urchins. We swim along rock walls carpeted with small pink strawberry anemones, where bulbous-faced ling cod and half a dozen other species of rockfish shoot from crevices, where decorator crabs, covered in red seaweed and green algae, camouflage themselves among the red-plated mollusks, white anemones, and purple ring top snails. Frisky sea lions dart through the surrounding waters, checking out the awkward bubble-blowing humans. The sun breaks through the surface cloud cover and suddenly the water is infused with cathedral light, giving the seals the spotlight attention they deserve. The majestic giant kelp stalks *(Macrocystis pyrifera)* have been called the Redwoods of the ocean, although they remind me more of Jack's beanstalk, growing up to 2 feet a day. Giant kelp not only creates a forestlike alternative to tropical reefs that justifies (if barely) cold-water diving but also adds uncounted zest to our daily lives. Harvested by giant lawn mower–like ships, the kelp's algin is used as a binder in some 70 household products, ranging from lipstick to ice cream. It is even used to give beer a longer-lasting head, a phenomenon we'll evaluate later at a Pacific Grove pub.

"This is a new frontier," Ed Cooper grins after our dive. "There are pinnacles on this coast that have never been dived. Out there in the canyon they're finding new species of life every week."

Back by the Coast Guard pier I spot a more familiar life-form. A sea otter is grooming itself, diving and resurfacing, swimming on its back, cheerfully tearing the legs off a freshly caught crab with its pointy little teeth.

The sea otter is the unofficial symbol of the Monterey Sanctuary and a major income earner for gift shops and gallery owners from Santa Cruz to Carmel. You can buy plush otter dolls, framed otter photographs, ceramic otter candy dishes, or $2000 cut-glass otters. This may be why few locals are willing to mention that these terminally cute and cuddly mammals are also voracious predators, eating up to 25 percent of their body weight every day, competing with commercial fishermen for octopus, crab, and urchin. They are also a species of marine weasel that is into rough sex. While effective for grooming their fine pelts or cracking shells against small stones they place on their bellies, the male otter's arms (legs, whatever) are simply too short for getting a good grip on a mate. So, the male gets firm purchase by biting down on the nose of the female before going for a little splendor in the kelp. Afterward you can often spot the females hauled up on rocks along the shore, their fur matted and their noses bloody. Breeding females are easily distinguished by the scars on and around their black nose leather. Knowing this, it is hard not to imagine that a female with a heavily scarred nose might get a reputation as an easy otter.

However you might feel about sea otters as role models for America's youth, we still owe these weasels big time. There used to be as many as a million corkscrewing through the coastal waters of the Pacific from the Russian Far East, along the Alaska coast, all the way down to Baja. But eighteenth- and nineteenth-century fur hunters decimated their population. In California the sea otter was thought to be extinct until 1938, when a raft of up to 300 was discovered living along the rugged coast of Big Sur. With the help of an environmental group called Friends of the Sea Otter and a Fish and Wildlife Service finding declaring the California otter a threatened subspecies, there was something of a rebound, with almost 2400 animals living along the coast by 1995. Beginning in the late 1990s, however, the otter population began an unexplained decline of about 5 percent a year. While suspected causes include the mammals losing out in their competition with commercial fishermen (who are active in the sanctuary), drowning in fishing traps and nets, being sickened by an unidentified toxin, or some combination of factors, no definitive cause had been established going into the fourth year of their population decline.

Beginning in 1988, more than half a dozen new sites were added to the national marine sanctuaries system. In Massachusetts it was oil drilling and the threat of sand and gravel mining on the Stellwagen Bank, a rich fish habitat and whale feeding area 25 nautical miles off Boston, that inspired a successful campaign to establish a sanctuary. Today, close to a million whale-watchers visit Stellwagen each year. It is also a major fishing ground for high-priced but increasingly rare bluefin tuna. Another sanctuary was established off the wild Olympic Mountains coast of Washington with its huge bird rookeries, cobbled beaches, pine-covered sea stacks and still-mysterious deep-ocean geothermal vents. The Flower Garden Banks Sanctuary 100 miles out in the Gulf of Mexico includes the northernmost reef systems in America and is home to endangered loggerhead sea turtles, manta rays, whale sharks, and spotted dolphins. Every August, at around 9:15 P.M. on the eighth day after the full moon, star corals start releasing smoky sperm while other parts of the reef seem to explode, releasing millions of tiny gametes, fertilized and unfertilized coral eggs, that mix with the sperm in the warm summer current. As if on signal, other simple creatures including sponges, tube worms, and brittle stars begin spawning in what is literally an orgy of life. Darting in amongst the predator fish feeding on the eggs are coral researchers with nets and bags, anxious to capture and study samples of this earliest stage of endangered coral reef life.

At other times of the year, East Texas charter boats come to Flower Garden for shark dives, when schools of hammerheads swarm the area in large numbers, along with the occasional 12- to 14-foot tiger shark.

At the Hawaiian Islands Humpback Whale Sanctuary, I have snorkeled with turtles and watched humpbacks breaching, leaping full-bodied out of the water as if their great winglike pectoral fins might actually lift them into the other blue domain.

I've also seen them at the other end of their migratory pattern, feeding and breaching off Point Adolphus in southeast Alaska. A mother whale slaps her 15-foot-long pectoral fin on the water as her baby swims nearby. A 45-foot whale breaches full out of the water, jumping eight or nine times right next to our ferry. No one really knows why they do this, although theories range from it being a way of removing parasites to "wouldn't you if you could?" Other whales swim close enough to the boat so that all you can hear is the "humphing" exhalation of their breaths and the snapping of camera shutters. Since the end of commercial whaling, humpbacks by the thousands follow this migratory pattern, spending their summers feeding in Alaska and their winters making babies in the protected waters of their sanctuary off Hawaii. That certainly convinces me these animals are highly intelligent.

America's 12 marine sanctuaries encompass 18,000 square miles of ocean, or approximately .05 percent (one-twentieth of 1 percent), of our EEZ frontier. California has a third of them (Channel Islands, Monterey Bay, Gulf of the Farallones, and Cordell Bank), while Alaska, covering more than half the U.S. coastline, has none. Alaska's congressional delegation of two senators and one representative does not want any more federal presence in their state. In fact, Representative Don Young once pulled out a buck knife during a heated debate with a congressman from New York whose advocacy of federal protection for Alaskan wilderness he did not approve of. He also cried when large sections of his state were set aside as parks and wilderness.

Throughout its short life the marine sanctuary system has been little known and poorly funded, although that is beginning to change. In 1991 the Potter Commission, an outside review panel commissioned by NOAA, suggested that $30 million a year would be "adequate" to support the sanctuaries. At the time they were receiving $4 million. In 2000 the sanctuaries budget was raised from $14 million to $26 million, with $35 million pledged for 2001, a significant increase; although, as Chris Ostrum, the diving coordinator for the sanctuaries, points out, after years of starvation budgets, "we soaked that up like a dry sponge in a puddle of water." In 2000 most of the sanctuaries had a manager but lacked outreach and science coordinators. Three sanctuaries did not even have their own boats. The National Park Service, by way of contrast, had a $1.4 billion budget in 2000.

While the sanctuaries may not need that level of funding, it is important to recall that going to the ocean remains the number one outdoor recreational activity for Americans. The average American spends 10 vacation days on the coast every year, with more visitors going to places like Jones Beach, the Jersey Shore, Puget Sound, and Hawaii than to Yellowstone or any other national park.

Aware of this popular preference, national park superintendents of seashore and water parks like Cape Cod, South Padre Island, Point Reyes, Glacier Bay, and St. John's in the Virgin Islands have begun meeting every year to discuss how to better service and protect their beaches, reefs, fisheries, and other marine resources for the oceangoing public.

At the same time, funding increases for NOAA's sanctuary program have come with a congressional caveat that no new marine sanctuaries be designated until the existing ones establish active programs of education, research, and onshore facilities.

"To say we can't have more because we aren't taking care of the ones we have is foolish," argues Sylvia Earle. "It's stupid to restrict new sites. Unless we begin to survey these places and give them protection now we lose them forever. We're using up our resources so fast, we have to act sooner rather than later." She pauses. "I don't want my grandchildren looking back and saying, 'You were there, and there were blue whales and coral reefs and you let them go,' and being ashamed of me—and they should be if we let this happen."

Like the national parks, each of the marine sanctuaries has its own unique history and story. Earle and photographer Wolcott Henry show and tell many of them in their 1999 *National Geographic* book, *Wild Ocean: America's Parks Under the Sea.*

In addition to writing and speaking out, Earle has become the sanctuaries' leading promoter through her 1998–2002 Sustainable Seas Expeditions, funded by grants from the California-based Goldman Foundation and *National Geographic*, with additional support, including staff and ships, provided by NOAA. Sustainable Seas uses a pair of manned (or more often "womanned") one-person submersibles to explore the deeper waters of the sanctuaries. Along with public education, it aims to identify key habitats, create sonar and visual images of various bottom types, compare fish populations inside and outside of no-take zones, study impacts of bottom trawling and other commercial fishing gear, and monitor runoff and water quality within the sanctuaries.

Bob Ballard's Jason Project is also increasingly active in the sanctuaries. Its live and archived Internet and broadcast feeds to classrooms across the nation have helped make science and ocean exploration real for millions of fourth- to ninth-grade students. Among its programs in 2000–2001 were interactive (and simultaneous) feeds from the Johnson Space Center in Texas and the Aquarius underwater habitat off Key Largo and from on and below the waters of the Humpback Whale Sanctuary in Hawaii.

The Farallones Marine Sanctuary is visited by whales including orcas, grays, sei, fin, sperm, humpbacks, and up to 80 blue whales, the largest creatures on earth. Its three small craggy islands, sometimes visible from the Golden Gate Bridge just 30 miles away, are a major nesting site for tens of thousands of seabirds, including storm petrels, common murres, and clownlike puffins, giving the area the feel of a mini-Galápagos. "San Francisco is a major city with a wilderness at its shore," notes sanctuary manager Ed Ueber.

On the rocky trails above its steep cliffs, South Island is littered with squawking, nesting birds. Its few small beaches are jammed with snorting elephant seals and barking sea lions that are using their larger cousins as sofas. But 100 yards offshore, floating face down in the cold saltwater, all is deathly still. Visibility in the Farallones gray-green waters is not very good, because of the nutrient upwellings from the deep ocean below. A strange sense of timelessness encompasses me as I stare at the shafts of surface light—when suddenly there's a rushing snout like a sea-launch missile, a flat black eye rolling back in its socket, a gaping jaw, and rows of razor sharp teeth that—bang! The video turns to electronic confetti as the camera is struck a stunning blow by a 17-foot, 3000-pound predator. Even reviewing the videotape, it's not reassuring. As a diver and body surfer, I wonder what drew the shark to the video camera attached to the bottom of a small floating surfboard when there was no bait or chum in the water. Shape recognition is the likely answer; surfboards look a lot like seals on the surface to certain blurry-visioned sharks.

Surfer Rob Williams was in the lineup off the north jetty in Humboldt Bay when he was misidentified. "I saw a set forming outside and began paddling toward it when the back end of my board pushed up," he recalls. "I looked back and just had a quick impression of a shark's mouth open and all these teeth, and it got a grip on me and the board and took us under. I was sort of sideways in its mouth. I felt it let up a little and then bite down again to get a better grip on my legs, and then it just started shaking me like a dog shakes the hell out of a toy. It must've lasted maybe 10 seconds. I thought I was dead. I was hitting it without effect and then I saw its eyeball rolled back and ran my hand up its nose and just jammed my thumb into its eyeball and held it. Then it spit me out and shook its head like it was irritated and took off fast—just boom—shot down into the darkness and I was there underwater alone. I popped up to the surface and climbed back onto my board. I thought I was okay but then looked back and saw this huge gaping wound in my thigh surrounded by bits of flesh and all this blood in the water and I started yelling and guys came and got me to shore."

After losing about a third of his blood on the way to the hospital, Williams was rushed into surgery where he was operated on for three hours. He survived his ordeal with 160 stitches, a dinged surfboard, and a nasty scar. Today, he again rides the waves off Humboldt, although a bit more cautiously. "The shark was in its element. I just had the bad luck to get nailed that day," he reflects. "I figure I'm still safer in the water than driving around in my car."

Farallones researcher Ken Goldman spent more than four years studying white sharks off South Island from a 17-foot Boston whaler, where he got a close-up feel for his subjects. He's convinced that on the rare occasions when these sharks do attack humans they quickly realize they have hit the wrong prey, which is why so few attacks off California (about 1 in 40) prove fatal. "I believe most of the time they will check out people in the water and realize they're not prey, and the people will never know the sharks are there below them." He means that to be reassuring.

Not a lot is known about *Carcharodon carcharias*, the great white shark. It can grow to more than 20 feet in length (the size of an average living room), weigh more than 3 tons, and rip 30-pound slabs of meat off its prey with a single scooping bite. Recent research, including dissection of dead animals, suggests the sharks pup off Baja and southern California. They produce litters of six to nine young that emerge from mama shark as formidable, toothsome, and independent-spirited 4- to 5-footers. They live anywhere from 35 to 70 years in the wild, spending their early years darting around the ocean gorging themselves on fish, squid, and smaller sharks. As they grow larger and less flexible, their diet and habits begin to change. At around 10 to 12 feet in length they reach sexual maturity, taking on a hard, rounded shape. Instead of chasing after fast-moving fish, they begin ambushing fat-laden, energy-rich pinnipeds including elephant seals and sea lions. White only on their bellies, the sharks are shaded gray and black above to camouflage them as they cruise rocky coastal bottoms stalking their prey from below.

Possibly the greatest concentration of white sharks in the world occurs every fall off the Farallones Islands during elephant seal breeding season, when as many as 40 sharks gather in an area about the size of Central Park.

"I get a call a month from people asking, 'Can I go out and dive with great whites?' I say I won't allow my people to do it, but *can* you? Yes, you can," Ed Ueber says, smiling wearily. Tanned and bald with a gray corona of hair, warm brown eyes etched with crow's-feet, tan chinos, and a checked flannel shirt, Ueber works out of a converted red-roofed Coast Guard station on the waterfront in San Francisco's Presidio. He and Billy Causey in Florida are the program's old salts, each having managed their sanctuaries for more than 10 years.

Ed served on navy submarines, in the merchant marine, as a shipwright on the *Morgan,* a tall ship in Connecticut's Mystic Seaport, and as a commercial fisherman on a dragger and lobster boat before joining the sanctuary program. Nowadays he roams his sanctuary from the top of the old Coast Guard lighthouse on South Island to 14,000 feet below the surface, where he traveled in a submersible trying to determine the fate of 50,000 barrels of nuclear and toxic waste the navy dumped there during the Cold War.

"The majority of the barrels are around 5000 feet down. In 1998 we had the British out there, with a towed wire that detects radiation. It got some readings about three times normal background, but nothing dramatic," he reports.

His other concerns include ship traffic in and out of San Francisco Bay and oil spills, including three in the late 1990s that killed more than 10,000 birds and tarred coastal beaches and estuaries. He also tracks plans for desalinization and cable laying, "strange effects" on birds and rockfish from years of warmer-than-normal waters linked to climate change, and the impact on Cordell Banks (a biologically rich, shallow seamont) from commercial trawling.

"The original agreements, when the two sanctuaries were established [in 1981 and 1989] was that they'd be open to fishing. So I'd like to keep my word to the fishermen," he explains. "You have to keep your word. But we also have to maintain good habitat as part of our requirements." He shrugs, knowing he sometimes faces contradictory responsibilities.

Outside his office window I watch a container ship from the Cho Yang Line pass by, also a ferry boat, a windsurfer, and a large sailboat with its orange and white spinnaker out. It's a water ballet on an urban estuary.

"People from San Francisco and Marin want us to protect their ocean," he says. "Not every sanctuary has that support. We also have the highest visitation of any sanctuary within the U.S. Ten million visitors a year to Stinson Beach, Muir Beach, Tomales Bay, Bodega Bay, a million of them are boaters."

"But how many know they're actually visiting a national marine sanctuary?" I wonder.

"Maybe a million know they're visiting a sanctuary," he guesses, "up from maybe 100,000 a decade ago."

While we're talking, his office manager hands him a FedEx package, which he promises to drop off before five o'clock. Ueber operates with a staff of three and a shoestring budget that finally topped $500,000 with the funding increases of 2000–2001. Like most of the sanctuaries, what really keeps his operation afloat is local commitment, including more than 250 volunteers for programs like Farallones Beach Watch, which trains coastal residents to testify in court on oil spills, respond to mammal strandings, give beach profiles on depositions of sand and debris, tally up birds and mammals, and watch and report on the natural cycles of the sea.

"Should we be keeping these areas in a natural state? Does the public want to preserve areas as marine wilderness? That's not the way the [sanctuary] statute reads," points out Brad Barr, long-time manager at the Stellwagen Sanctuary and now the program's senior policy analyst, based out of Woods Hole, Massachusetts.

"Look at the response of the public to drilling for oil in the Arctic Wildlife Refuge," he suggests. "We value that. Isn't there a similar value in the marine environment? It's important for the public to weigh in. Do you want that wilderness or not? Because every area can be exploited now. There's deep water corals in the Gulf of Maine being destroyed by fishing gear, unique unstudied ecosystems, where now you have fishermen prospecting for new fisheries."

Although he says he left Stellwagen after six and a half years to make way for new blood, more than half a dozen knowledgeable people I spoke with suggested Barr was kicked upstairs for having pushed too hard for the establishment of marine protected areas, or no-take zones, in a New England sanctuary where a single bluefin tuna can earn a fisherman $20,000.

Still, he is hopeful that change is not only possible but likely. "We can use these sanctuaries to develop models for ocean governance. We could be a great test bed for effective management and resource protection of our oceans if that's what the public wants."

What the people of Sapelo Island off Georgia mainly want is to be left alone. Gray's Reef Sanctuary is 17 miles off Sapelo. Its 50- to 70-foot-deep hard bottom is said to make for some fine diving, with lots of grouper, angelfish, sea bass, big sponges, and jelly-munching turtles, but I did not get to dive there. Instead, I spent a windy, blown-out day on this wild barrier island, which is also a NOAA estuarine research reserve.

I arrive at Sapelo with friends and fellow alumni of the Institute for Journalism and Natural Resources, a kind of continuing education program for environmental reporters. We approach the island by boat, through oxbows of tidal marshes and the open waters of Doboy Sound, where we're accompanied by a couple of curious bottlenose dolphins. Pulling up to the dock, I'm distracted by a line of shorebirds standing atop the first natural oyster reef I've ever seen. It stretches like a whitewashed garden wall 150 yards along the marsh's green-yellow edge. Onshore we're greeted by Maurice Bailey, a muscular, broad-cheeked man in his thirties with an easygoing manner, a brown felt hat, and carved bone hat band. Once aboard his yellow school bus and headed down a wide sandy road through humid low country, he begins to give us the basics: Sapelo Island is 11 miles by 3 or 4 wide, 17,000 acres altogether, with a marine lab, post office, and daily ferry.

One of the Sea Islands, Sapelo was first settled by Indians, then Spanish, British, and Americans whose West African slave laborers managed to retain much of their Gullah and Geechee languages and culture. The Africans, producing sugar, cotton, and indigo for white plantation owners proved more resistant to malaria and other coastal diseases than did their Anglo kidnappers. As a result, following the Civil War, many freed blacks chose to stay on along the coast, turning to fishing, crabbing, and oyster dredging as their primary means of livelihood. At the same time wealthy robber barons like the Carnegies and Vanderbilts began acquiring many of the offshore islands as private holiday camps. Today, the mixed Gullah and Geechee culture continues to exist along the Sea Islands of South Carolina and Georgia despite increased land pressure from a new breed of federally insured wealthy white developers.

Maurice Bailey drives us past "The Big House," a pink and white mansion with 14 bedrooms, a columned entryway, rows of live oaks draped in Spanish moss, and its own bowling alley. Built by cotton magnate Thomas Spalding, it was later owned by Hudson motor car executive Howard Coffin, who then sold it to tobacco heir Richard (R. J.) Reynolds in 1949. Today it is a guesthouse, rented out by the Georgia Department of Natural Resources.

We drive on to Nanny Goat Beach, where we're greeted by a couple of blue herons and Cornelia Bailey, Maurice's mother. She is a thickset gray-haired woman with shell earrings, wind-protective boat pants, and a feminist T-shirt, something about sisters doing their thing.

"We're all Gullah-Geechee," she explains to us. "But there's saltwater and freshwater Geechee. Sapelo people are saltwater. There was 350 or so on the island a generation ago, now just about 70. We used to have five communities on the island. I was born in Bell Marsh, and moved to Hog Hammock, which isn't so good. It floods there. But the better land was taken by Reynolds. Now we're holding on for dear life. Other islands got ruined. We went to this meeting with these businesspeople on how to encourage black tourism and we said, 'When you buy our land and move us off the land you're losing the culture that will attract these people. Black people don't want to come to these islands to see white people in $250,000 homes playing golf.' Now we are trying to hold on to what we have and we're all Baptists, so we're praying." She reflects on that a moment. "Well, one woman became a Jehovah's Witness. But she was freshwater, and don't know better.

"Writers come here and say we never leave the island." She gives me and my fellow reporters a baleful look. "Growing up when I did everyone had a rowboat and we always left to the mainland. We left, just not to live elsewhere. We didn't choose to live nowhere else. Years back we knew when to cut marsh grass for our horses and cattle and when to gather oysters and how to reseed the beds and when to fish with a drop line and hunt and when to leave them alone. We didn't call it an economy but it was, and maybe now we're just trying to hold on 'cause without the land and the water we have no reason to be here."

Today the Gullah-Geechee future looks bleak, she admits. "We need a miracle and fast, otherwise without an economic base we're doomed. We have 21 kids left here is all, and they go to school on the other side [the mainland]."

"How about tourism?" someone suggests.

"Tourism? Well, everyone don't want to talk as much as I do. We use [the Geechee] language as much as possible to keep it alive. And we still have some crafts, like making castanets."

Does she see the island and the coast staying the way they are?

"If the DNR [Department of Natural Resources] and other folks keep the rules, I don't think we'll be like neighboring states. If we keep or make tougher rules I think we could keep the Georgia coast like it is, without houses on the beaches and marshes being filled in." She tells us she's writing a book, *God, Dr. Buzzard, and the Balitto Man,* and goes on to explain the title.

"God is number one and you can't ask God for revenge 'cause preacher says turn the other cheek, so you go to voodoo, Dr. Buzzard, for revenge. Also, you can't ask God for money so you play Balitto, the numbers what the Cubans call it. Ten cents can get you twenty-six dollars so you gamble that way. Certain things you just don't ask God for."

I wonder if you ask God for a sustainable economy as I walk down the windswept beach where 200- and 300-pound Loggerhead turtles come to lay their eggs every year. It's beautiful in a bleak, wild way, with big blown-out white-capped waves cresting over the gray-green water. A pelican beats against the wind, barely moving. I walk along the foamy edge of the sea collecting whelk shells, past the bodies of small horseshoe crabs, the lipstick-red arms of female blue crabs, a female CNN reporter turning her back to the wind, clam and scallop and angel shells, and several large pieces of palm trunk driftwood. White grains of sand whip across the beach in ghostly slithering sheets, 6 inches off the deck. After an hour of being sand-blasted, I head inland, spotting a whitetail deer running through the sandy brush and sea oats in the quiet air just beyond the beach's protective 7-foot dunes. I walk along a trail of long leaf pine and Spanish bayonet, wax myrtle, saw palmetto, and southern magnolia, running into two friends, Susan and James, a radio reporter from Maine and newspaper man from North Carolina. We wander through a salt marsh under overhanging Spanish moss that lets out by the mansion. We walk over a pair of decaying wooden bridges arching a green scum-covered pond that take us to small overgrown islands. The second one has a Roman trellis and picnic area with inlaid tile flooring. We smile, imagining romantic jazz age picnics interrupted by the bellowing of alligators. Sapelo seems to have it all—an upland maritime forest, tidal salt marshes, oak and cedar hammocks, dunes and beaches, Indian shell middens, and not a single PGA-quality professional golf course.

That evening we're invited to the cinderblock Hog Hammock community center for a dinner of barbecued pork, corn bread, greens, smoked mullet, and other good-tasting vittles.

Afterward, out by the porch an older gentleman named Ben Hall of the Sapelo Island Cultural and Revitalization Society sits in a rocking chair, telling us of plans for an auto garage and youth center and some more history. "In 1834 this community was founded at the time of slavery when it was used for raising hogs. My great, great grandfather, Samuel Hogg, took the name from his job, but others took the name Smith from his wife's side, so Hoggs and Smiths are all the same family. White owners abandoned the islands after emancipation. We were more resistant to malaria mosquitoes so we stayed. In the 1950s the neighboring communities were encouraged to consolidate."

"Ben's being a little too nice," Maurice interrupts politely. "The land was stolen. People were pressured to leave. Richard Reynolds made everyone move here to Hog Hammock or to the mainland. Electricity only went to Reynolds's supporters. Electricity for everyone didn't come till 1965. We had one phone booth for the whole island till 1980."

A white lawyer working for the island residents mentions how Raccoon Bluffs, one of the lost communities, is on dry, beautiful upland. Shell Hammock is now the site of the University of Georgia Marine Lab. Bell Marsh and Moses Hammock are state campgrounds for hunters. People agreed to move but without giving up their titles, he claims. Ownership deeds go back 130 years on Raccoon Bluffs. But after nine generations there are also confusing deeds, false deeds, and missing wills. A court case will take time and money. At the same time, the Revitalization Society receives funding from the Reynolds Foundation, which complicates things further. Still, a number of residents believe that if the historic injustice of their displacement is righted, saltwater Geechee who have gone "to the other side" might return.

"We know of and stay in touch with 325 to 330 of our people on the mainland," Ben Hall tells us.

"People leave but stay near the water," Cornelia explains. "We have folks as far as on the west coast, but no one in the middle 'cause we're saltwater."

Fishermen are also "saltwater," as are dedicated surfers, professional sailors, and people whose souls just cannot find peace away from the shore. From Sapelo Island to Seattle, Laguna Beach to Cape Cod, Massachusetts, people struggle to maintain their connections to the sea, often for reasons they cannot easily explain, reasons as deep as the ocean itself. That may be why, despite their many financial, political, and environmental conflicts, our national marine sanctuaries have something going for them that many better-funded government programs lack.

Marine sanctuaries act as social and cultural magnets, attracting the support of citizen activists and coastal communities that have committed themselves to the long-term conservation, restoration, and preservation of America's Blue Frontier.

"I get calls all the time from people saying, 'What are you up to in our sanctuary?'" Ed Ueber tells me. "That feeling people have that it's their sanctuary, that they own it, and if the government wants to help them out fine, but it's really there for them and their kids—that makes this program work."

It is that same commitment, organized at the grassroots or seaweed level, that may yet save our final physical frontier from being pacified, tamed, or degraded.

Chapter Twelve

The Seaweed Rebellion

> To stand at the edge of the sea . . . is to have knowledge of things that are as eternal as any earthly life can be.
>
> —RACHEL CARSON, AT THE EDGE OF THE SEA

> If you like to eat seafood or swim in the ocean, it's time to get involved.
>
> —JULIE EVANS-BRUMM, FRIENDS OF LONG ISLAND SOUND

My cell phone rings as we're hunting for crocodiles, so I turn it off. I'm with Steve Klett of the U.S. Fish and Wildlife Service, the manager and sole employee of the 6700-acre Crocodile Lake National Wildlife Refuge, which shares north Key Largo with the upscale Ocean Reef Club. Along with perhaps 100 of the estimated 800 surviving American crocodiles, the refuge is home to the endangered Shaus swallowtail butterfly, Key Largo cotton mouse, wood rat, Eastern indigo snake, and Stock Island tree snail. Rosette spoonbills, sea turtles, manatees, bald eagles, and peregrine falcons also fly, paddle, and slither through this island swamp.

Steve is a rangy ranger, with receding reddish hair, sun-reddened skin, and wary blue eyes that let you know he's a man who doesn't mind keeping his own company. His office is in a cheaply built, wooden stilt house on a lifeless canal in an old, failed swamp development that was dredged without a permit. Still, it is affordable on Steve's $10,000-a-year refuge budget.

Right now Steve is in his Spartan second-story office writing a proposal for a rescue and salvage operation for the Stock Island tree snail. He shows me one of the snail's shells. It's gorgeous—about 3 inches long, fluted, with a brown and white swirl in it, like vanilla fudge ice cream. He hopes to relocate some of the last few hundred snails back onto the refuge. At present, they exist here in a housing subdivision in Key Largo and in trees around the parking lot at Monkey Jungle, a once-popular roadside attraction on the mainland. Predictably, their numbers are declining at these last two sites.

The refuge and adjacent state park, which include 11 miles of hardwood hammocks and mangrove swamps, are the result of a 1980s lawsuit against Monroe County in the Florida Keys. The suit was brought by the Audubon Society and local environmentalists organized as the Key Largo Citizens Coalition. They charged the county with failing to protect endangered species like the Florida panther, marsh rabbit, and manatee. As part of a settlement the county wrote a habitat conservation plan that led to federal and state protection for the undeveloped half of north Key Largo.

"The real problem is we've never determined what we want the Keys to look like," Steve reflects. "We've slowed growth, not stopped it, so it'll take twice as long to reach apocalypse. But if you say no growth around here, you get a huge backlash from the real estate and tourist industries, the guys with the money and the politicians in their pockets.

"The Keys were founded by smugglers and there's an attitude of taking for yourself and not giving back," he continues. "Like I tried to stop this development on Lower Sugarloaf Key. It was a house on a beach berm that threatened habitat for the marsh rabbit. It was going to be set between a hardwood hammock and a mangrove wetland. The cumulative effect of this one rich person's house would be awful but you can only stop a project if you can prove that specific development will destroy the species. By the time you get to that point, it's usually too late. That species will go extinct."

"So bye-bye bunny?"

"The world's changing too quickly," he grouses, sounding older than his 45 years, which is what can happen when you're sitting in the middle of a "hot spot" during a global extinction crisis. "When I lived here in the early eighties you never saw algae growth on the corals," he says. "No one had air conditioning. You had screened open-air porches and Bahama fans. Now you have these video stores leaving their doors open because it's too cold inside. We had a nesting colony of frigate birds on the Marquesas Keys that no longer exists, too many people on Jet Skis checking them out. At least here on the refuge I can see what I've done for the resource in my own little world."

I ask him to show me some of that world. He puts on his Oakley sunglasses, I my Ray-Bans, and we go down the outdoor stairs, across the sun-blasted Florida crabgrass and around to his truck parked in the warm shade under the house.

"We have nearly a hundred species of native trees and shrubs," he tells me. "Eighty percent of the plants are of West Indian origin: mahogany, gumbo limbo, paradise tree, poisonwood, and lignum vitaes so dense it won't float. The plants mostly came north on the currents or [as seeds] with birds, whereas the animals all migrated south from the mainland.

"There were some twenty developments planned for this area before the refuge," he continues as we drive past an abandoned trailer park. "Fifty thousand people were going to live here. There was Port Bougainvillea. They were digging lakes and ocean outlets for it before it failed in the eighties. Also, the Harrison tract, another failed development where crocodiles are now laying eggs along the banks of the canals they dug."

We drive up Route 905 past an old dump and Cold War Nike missile site where turkey vultures squat sentinel-like on the abandoned observation deck rails, as if listening for the sound of approaching Mig-21s out of Cuba. We drive past the old cock fighting arena and another trailer park that was wiped out in a tropical storm.

We pull off the two-lane "highway," park, and walk down an abandoned development road where green jungle growth is pushing up through the crumbling asphalt. From here he leads me through a dense thicket of vines into a hardwood hammock where the forest floor opens up under a native bamboo and poisonwood canopy and clouds of blood-hungry mosquitoes begin whining around us. "We'll use this if we really need to," he says, showing me a can of Deet-based bug spray before putting it back in his pocket. A mosquito lands on my upper lip. I tap it with my finger, which comes away with a smear of blood. A dozen more land on my arms, neck, and ears. He points out wild coffee and pigeon plum and shows me where raccoons have marked the bark with their claws. I put on a thin shell jacket for some protection. He points out a gumbo-limbo tree. "They call it tourist tree here because of its peeling bark, like a sunburned tourist." I notice the mosquitoes hardly touch him, probably bled him dry years ago.

"The tree snails feed on lichen and moss, although I can't see any right now." Steve's moving carefully among the trees looking for snails while I'm doing a little shudder dance trying to keep the bugs off. The heat is, of course, stifling. Steve talks about invasive versus native species. Human-introduced fire ants, feral cats, and black rats have become a problem.

"Would you like some of this?" he finally offers as he notices my furious slapping. "Somehow they don't bother me so much as they used to. Maybe I've developed an antibody." I snatch the aerosol can, spray my hands until they're wet, and rub the liquid toxin all over my face and arms. It begins to work immediately. They don't bite, they don't even light.

Back along the abandoned roadside he shows me where volunteers from the University of Florida have been helping him plant wild lime, a favorite food of the swallowtail butterfly. I spot several small bright butterflies but none of them are swallowtails.

We drive to the Harrison tract and hike along its brown-water canals to where there's a dug-out depression and drag marks from a mother croc.

"This is where we found 16 eggs that had been opened up, and two empty shells where she may have gotten live young out." I start to scan the canal's murky water for shy, toothy green lizards. I can see the croc's most recent path, where the shallow tail drag marks continue over loose limestone fill.

"They average 7 feet around here, though I've seen 10-footers," he says.

We walk along the overgrown canals for awhile, drinking up our water and sweating through our clothes but encountering nothing more threatening than the slither marks of snakes, butterflies, and bird song.

"I know where else we might see one," he offers.

We're standing on top of guard rail posts on the shoulder of 905 by Barnes Sound as Cadillacs, Monteros, and a white stretch limo from Ocean Reef zip past. There is a lot of mangrove swamp beyond the roadside hurricane fence but no crocodiles. "Hot part of the day they tend to hang in the shade of the mangrove roots," he explains as we scan the water's edge with our binoculars. There are lots of egrets out and about fishing the shallows, so at least I can enjoy watching them earn their living. We go up to the toll bridge where the sound rolls out to Florida Bay and the gulf. "Sometimes they take to the saltwater," he says, but not today.

The Turkey Point nuclear power plant just across Card Sound has 80 to 90 miles of cooling canals with wide levees where American crocodiles are thriving. "A biologist there digs out small freshwater ponds on the levees using a backhoe," Steve tells me. "They like freshwater as infants. They have up to 12 nests over there. It's odd to see this technological monolith and these primitive creatures in proximity to each other like that."

The same could be said for the Crocodile Lake Refuge with its primordial coastal critters and the Ocean Reef Club with its artfully created illusion of a manicured tropical paradise. Florida is, of course, famous for its wild alligators tramping across its emerald green golf courses, as well as for its pleasure boats running over and mutilating slow-moving manatee sea cows.

I ask Steve if he's involved with any ocean protection or environmental groups.

"No," he says, unsmiling. "But when people ask me what environmental group I think they should contribute to, I say Planned Parenthood."

For all their misanthropic tendencies I find people like Steve Klett oddly inspiring. Maybe because they still care.

All across coastal America people who care are beginning to act on their beliefs. The number of blue groups is growing, along with public awareness about our oceans. Major activist organizations include the Ocean Conservancy, which sponsors the National Beach Cleanup Day every year and is headed by a former Coast Guard admiral; the Surfrider Foundation; the Cousteau Society; the American Oceans Campaign, founded by Ted Danson, the actor and ocean activist; Coast Alliance; Seaweb; and the Water Keepers, made up of more than 40 boat-based operatives who work to protect rivers, bays, and estuaries from coastal Georgia to Cook Inlet, Alaska. Robert F. Kennedy, Jr., is their president. There are coalitions of groups like Restore America's Estuaries, the Marine Fish Conservation Network, Clean-Water Network, and Ocean Wildlife Campaign. There are more regionally based groups including the Littoral Society, Clean Ocean Action, Reef Relief, and REEF, which cosponsors the Great American Fish Count every June during which divers help scientists census local fish populations.

There is the influential Chesapeake Bay Foundation and venerable Save San Francisco Bay, the North Carolina Coastal Federation, Friends of Long Island Sound, Tampa Baywatch, People for Puget Sound, and the Gulf Restoration Network, among others. There are a number of groups that focus on specific marine wildlife such as the Pelagic Shark Foundation, Save Our Wild Salmon, Friends of the Sea Otter, Save the Manatee Club (cofounded by Jimmy Buffett), and the Sea Turtle Restoration Project.

Hundreds of local organizations are also making their presence felt through constructive engagement with fellow citizens, resource users, and government agencies. In Santa Cruz, California, for example, Save Our Shores, a group that was founded to protest offshore oil drilling, has evolved into a citizen-watchdog and resource for the Monterey Bay Sanctuary.

Concern over coastal sprawl is leading to the creation of countless new groups like SAND (Seeking a New Direction), a recent merger of Gulfport and Biloxi, Mississippi, neighborhood groups opposed to the proliferation of malls, gravel pits, condo towers, and casinos along their once scenic shoreline.

Green groups like the San Diego Environmental Health Coalition are also increasingly taking on maritime causes, in their case problems related to the home-porting of navy nuclear aircraft carriers.

The environmental group Greenpeace first gained a global reputation when its members used inflatable Zodiacs to block the harpoon guns of whaling ships. Today, Greenpeace continues to maintain a strong interest in marine issues like overfishing, calling for a ban on factory trawlers. The more militant Sea Shepherd Society, founded by Greenpeace dropout Paul Watson, has gained a Corsair-like reputation for using its ships to ram pirate whalers and drift-net vessels.

In the 1990s mainstream environmental groups including the Aubudon Society, Sierra Club, the Natural Resources Defense Council, National Wildlife Federation, Earthjustice, and Environmental Defense also began focusing on the Blue Frontier. The San Francisco–based Earth Island Institute, along with its efforts to return the orca Keiko (screen name Willy) to the wild, led the successful tuna boycott that resulted in dolphin-safe labeling. One of its spin-offs is now working on turtle-safe labeling for shrimp harvested using turtle excluder devices (TEDs). Both labeling programs have been challenged by foreign fishing nations (and Washington politicians), arguing that requiring fishing boats to use gear that saves marine wildlife is an infringement on free trade. The World Trade Organization (WTO) even ruled against the U.S. requirement that imported shrimp be caught with TEDs. That is why hundreds of the protesters at the 1999 Seattle WTO demonstrations were dressed as sea turtles.

Despite resistance from free-trade absolutists, however, the idea of labeling sustainable seafood has become increasingly popular. In 1999 both the Audubon Society and the Monterey Bay Aquarium put out consumer guides to seafood, suggesting that people might choose to eat wild Alaskan salmon, for example, rather than farmed salmon, whose wastes can pollute surrounding waters; or choose to order abundant species like calamari, mahimahi, and striped bass, rather than overfished species such as Chilean sea bass, scallops, and swordfish.

In 2000 this labeling effort expanded under the sponsorship of the Marine Stewardship Council, set up by the World Wildlife Fund and the Dutch-based multinational Unilever, one of the largest commercial buyers of fish in the world (Gorton's Seafood is one of their subsidiaries). Among the first human prey items to win the council's "Fish Forever" seal (certified sustainable by independent experts) were West Australian rock lobster and Alaskan wild salmon. Among the first U.S. companies to use the Fish Forever seal was the Boston-based Legal Seafood restaurant chain, which buys 100 tons of fish and shellfish every week.

"We've been in the fish business for fifty years, and I'm interested in being in the business another fifty years," explains Legal Seafood's CEO Roger Berkowitz. "The only way we can do this is to participate in some kind of conservation effort. What I like about this program is that it's a positive approach that can help motivate people. It's a way we can help educate our consumers and also encourage fisheries to sustain themselves."

Just a few miles south of Crocodile Lake I run into folks from the local Marine Mammal Conservancy, who have decided to sustain and restore the Blue Frontier one fish at a time. Before I know it, I'm one of 16 people, thigh deep in clear tropical water, moving Sheri and Florence around in a circle. Sheri is 5 feet and Florence is 5 feet 3 inches, and they're both a little disoriented after a long day's flight on American Airlines from Chicago and the drive down from Miami, followed by cameras and reporters to this small cove near Captain Slate's Atlantis Dive Center. The cove has been enclosed with orange plastic fencing and iron rebars so the girls can spend a safe night getting used to the warm 82-degree saltwater. They've spent the last 18 hours in aerated, water-filled giant coolers.

Sheri and Florence, if you haven't yet guessed, are nurse sharks. Captured as little nippers, they grew up in the Seashell Pet Shop in Chicago, where they eventually grew too large for their pool. That's when Wendy Rhodes, a pale, blonde, animal rights activist from Los Angeles, found them and decided to launch a rescue mission. She made a call to Rick Trout, an ex-navy dolphin trainer who now lives in an old Keys cottage surrounded by cats, coconut palms, and bougainvillea, works as a commercial diver, and helps coordinate marine mammal rescues. Rick organized the planned shark release from this end and is now pushing Florence in my direction. She seems to be losing a little of her jet-lag lethargy. Her skin, as I guide her through the water, feels more like raw silk than sandpaper, and I can feel her muscles beginning to work beneath it. She tries to turn away from the circle, but I firmly direct her back along the line. Shark wrangling could seem macho, except for the fact that Sky, a 7-year-old blonde pixie in a blue wetsuit, is now hugging Sheri. "No hugging the shark, you have to pass her along," her mother gently chides. There are five kids, ages 7 to 12, carefully dispersed in the circle, including two preteen brothers, Brady and Brandon. "Hey, have you heard of fish and chips?" Brady asks one of the sharks he's pushing along, trying to get a response from his brother, who ignores him. A video crew from Miami's Channel 7 is wading around the edges of the circle, including a young Latina reporter who keeps trying to get control of her hair in the evening breeze. Their lead news tease tonight will be, "Sharks gone wild in south Florida."

Soon the sharks are swimming on their own, and we quickly leave them to their adjustment. Back on shore I talk to Clifford Glade, a medical doctor and veterinarian here with his young daughter, Nikki. The day before, he had set the legs on two rare cranes who had broken them flying into newly constructed electrical towers.

I mention how this release may only be a net gain of one shark for the reef. Earlier in the day, while on patrol with sanctuary cop Dave

McDaniel, we'd spotted a cabin cruiser named *Sea E Oh* with three guys cheering on a woman companion who was hauling in a large nurse shark on rod and reel.

"You have to start somewhere," Dr. Glade says. "I'm 48. Nikki is 10. I want her to grow up where there are still sharks and wild birds and a coral reef she can enjoy."

The next morning the sharks are in black tubs on the fantail of the 42-foot *Coral Princess* as we head out to Elbow Reefs and the wreck of the USS *City of Washington*, the nineteenth-century navy ship that returned the bodies of the battleship *Maine*'s dead from Havana harbor at the beginning of the Spanish-American War.

The young woman reporter from Channel 7 is back with her crew, interviewing Rick. "Hopefully, this will show people that this is where sharks belong and keeping them as pets is just a bad habit that they should get over," he says. We arrive at the reef.

Captain Spencer Slate is a fair-sized, mustachioed good ole boy and a newlywed (he just married Sky's mom). He suggested this release site because he brings scuba divers here every week, which may provide the sharks some level of protection from fishermen. He and Rick have also stapled small ribbon tags onto them for future identification (the sharks did not seem to mind, although Wendy went a shade paler than normal). There's a barracuda on the reef named Lightning and an eel named Perry. On his dives Slate holds a fish in his mouth, that Lightning the barracuda will then grab. So far the barracuda has been able to distinguish the fish from Spencer's masked nose.

We dive down to the shallow wreck lying in 25 feet of water. There are some nice fan corals growing on its iron ribs, along with free-swimming wrasses, sergeant majors, parrot fish, and lobsters in rusted out holes and rock crevices nearby. After everyone's settled on the bottom, Rick and Captain Slate swim the two sharks down. Slate tries to feed pieces of squid to them but they spit it back out. Suddenly, Perry, the 6-foot green moray eel swims out from under the wreck and slides over Slate's shoulder to grab some of the squid, catching him by surprise.

This is a long way from the Sea Shell Pet Shop, and the two nurse sharks don't look too happy, lying on the bottom together, one with its pectoral fin over the other, as if to comfort it and say, "Where the hell are we? Where are the goldfish, hamsters, and parakeets?" There are now 17 divers and two sharks on the bottom, plus a small local nurse shark who's willing to be fed if they aren't. Half a dozen water-sealed cameras are recording the event in a kind of media feeding frenzy that will replay on Miami's TV stations this evening.

Of course, it is going to take more than a few news stories to educate the public about marine wildlife, especially with so many people swimming through the shallow end of the gene pool. A few days after the nurse shark release, a tourist from St. Petersburg boating off Key Colony Beach spots a bunch of fins in the water. Deciding he's going to go swimming with wild dolphins, he jumps into the middle of a school of aggressive 7-foot bull sharks, one of which bites him in the foot.

I call Rick six months later to see how the nurse sharks are doing. They're now feeding themselves, he tells me. And they moved off the wreck to Finger Reef, where they've been seen in 50 feet of water. Wendy has also located and shipped three more nurse sharks from a Pizza A Go Go in San Jose, California.

In Key West I reconnect with a good-looking, sun-etched couple named Craig and DeeVon Quirolo. Several years ago I produced a PBS report on the group they founded, Reef Relief, which now has some 5000 members, with an equal number turning out for their annual Cayo Caribbean Music Festival (and fund-raiser) by the old fort in Key West.

Craig, a former charter boat captain, and DeeVon, who used to produce illustrated tour guides, started designing and placing mooring buoys in 1986 so that dive boats would not drop their anchors on live coral. They also began to educate divers and snorkelers not to touch the coral (which can remove a protective layer of slime and expose the polyps to infection). They've worked to improve water quality in the Keys (which led to a local ban on phosphates) and helped establish the national marine sanctuary. Today, through Reef Relief, they continue to install dive buoys and promote marine parks in the Bahamas, Jamaica, Cuba, and other parts of the Caribbean.

"For years we worked mainly on a grassroots level," Craig tells a Reef Awareness Week dinner at the Pier House. The locally catered event has drawn a tanned, rum-friendly crowd made up of a cross-section of Keys society: a schoolteacher, radio disc jockey, dive-charter operator, motel manager, supermarket manager, commercial fisherman, nature guide, state biologist, graphic artist, and some 70 other movers, shakers, and swimmers. Corporate and government sponsors include American Express and the South Florida Water District.

"After working here in the Keys, we decided to go to D.C. and lobby for a sanctuary, which we got," Craig tells them. "We figured the government would get involved and save the reef, only it hasn't. Then we thought science would save the reef, only there's all this disagreement among the scientists. So now we're back to saying it's up to us to save it. We can't expect the government or the scientists to save our reef for us. We're going to have to do it ourselves, by educating young people and reaching out to people in other parts of the nation, and the world, and by telling them about this living treasure we've got down here."

Dauphin Island, Alabama, reminds me of Key West when I was a kid, a kind of laid-back island without a lot of commercial distractions from the magic of its open sky and rainbow-streaked waters. George Crozier, the director of the Dauphin Island Sea Lab is talking to me and Ray Vaughan about nutrient runoff and its possible link to blooms of stinging jellyfish in the Mississippi Sound and Mobile Bay. "Hey," Ray grins. "We'll tell the rich folks on the beach that if they bring in more hog farms, you'll never swim again."

George laughs, "That's a Wildlaw statement for ya."

Ray, 6 feet 4 inches, 235 pounds, and tough as a barracuda, is a "'bama boy," born and raised with a taste for fishing, barbecue, and guns. He's also Alabama's leading environmental lawyer and founder of the nonprofit Wildlaw Foundation in Montgomery. He and his buddies, fellow lawyer Ned Mudd and trading post operator and trapper Lamar Marshall, have filed about 95 percent of the state's environmental lawsuits, gaining themselves a reputation as among the South's preeminent eco-rednecks. Ray recently forced the Department of Environmental Management to hold public hearings before issuing permits for corporate hog farms to operate along the state's waterways. "Alabamans aren't stupid," he says. "We can see what's going on in North Carolina," where nutrient runoff from hog factory waste has created biological dead zones along the coast.

Dr. George Crozier is older than Ray, with blond hair turning white, a craggy, sun-reddened face, and a fun-loving, hyperkinetic style not often found among the more staid, northern breed of scientist. "Our lab started in 1971 on a mosquito, bug-infested peninsula," he tells me. "Now we're on a mosquito, bug-infested barrier island."

Fourteen miles by 1.5 miles at its thickest, with some 2000 winter residents and as many as 15,000 summer visitors, Dauphin Island has been repeatedly hit and reshaped by tropical hurricanes, including Frederick in 1979 that took out the old bridge to the mainland and in 1997 and 1998 by Danny and George, respectively, which destroyed a number of houses, some of which floated into each other. Hurricane George also moved sand dunes around and stripped away hundreds of feet of beach on the seaward side of the island.

From the water the island's narrow west end looks like a forest of wooden stilts on top of which several hundred houses have been temporarily secured. You could fish off the decks or out the bathroom windows of many of them where the storm-eroded sand has retreated underneath their pilings.

After Hurricane George, FEMA agreed to spend millions of tax dollars to protect the single road out here, but without requiring any additional public access to what's left of the beach.

"I'll be damned if public money should be spent for these owners to be making more money than they already do with their summer rentals, and with no benefit to the public," Crozier declares as we rock in a windy chop 100 yards offshore in one of the lab's 26-foot research vessels.

Later I go walking down the west end with Ray, whose mother used to bring him to Dauphin Island when he was a child. Now he's 38 with three kids of his own that he likes to bring out here. Just past the last stilt house, a rental unit called Tale's End, with waves rolling in under its pilings, the road ends at a barbed-wire-topped hurricane fence with a sign saying "Private Property." Much of the island beyond this point is little more than an 8-mile-long sandbar that's been washed over by the sea at least half the time during the past 50 years. It takes two minutes to walk its width from the ocean to the Mississippi Sound. Nonetheless, the owner tried to have it exempted from its federal CBRA (Coastal Barriers Resources Act) listing in order to qualify it for government subsidies. He had hoped to get FEMA flood insurance so people would be willing to buy $200,000 home sites (200 of them) on 1.5 miles of this narrow sandspit. He would have raised the money needed to install roads, sewers, and electrical lines by selling the rest of the spit to the state as a park for $20 million, a figure far above its assessed value.

Working with a group of local residents called Forever Dauphin Island, Ray sued the state and managed to kill the deal. "It was insane to build here," he says pointing to the west-end houses behind us. "But it would have been more insane to build further out."

Ironically, one of the land's owners, Riley Boykin Smith, is also Alabama's state commissioner of conservation and natural resources.

Surfrider is holding another Clean Water Paddle Out, this time at San Francisco's Aquatic Park. Surprisingly, despite Mark Twain's observation that the coldest winter he ever spent was a summer in San Francisco, the sun is shining without even a wisp of fog on the water. The bay's water temperature is, however, a brisk 57 degrees. Although the Bay chapter has 1000 members, only about 50 brave hearts have shown up this morning on the walled beach below the Ghirardelli chocolate factory. Todd Walsh, the slim, unshaven chapter chair, recommends a 4-millimeter wetsuit for getting out in the water.

A Beach Boys tape begins to play on the sound system as the small, wetsuited crowd hits the water, paddles out among the moored sailboats, and forms a circle in the water for a big whoop before moving past the breakwater and out into the bay. After about 15 minutes paddling around some buoys, they begin straggling back ashore. People cheer as they hit the sand. "Great." "Cool." "I'm stoked." "I saw a seal."

"It's a good cause and a good excuse to get in the water on a nice day like this," explains Becca Wheeler, a bright-eyed San Francisco nursing student who recently moved here from the leeward side of Oahu.

"How does mainland surf compare to Hawaii?" I ask her.

"This surf sucks," she says with a brilliant smile, stripping off her wetsuit. She likes the long rides at Waikiki. Here the rides are shorter and there are no channels out. Also, 57 degrees is the big chill compared to Hawaii's warm, gin-clear waters. I start to get misty-eyed talking to this pretty wahine, and realizing how long it's been since I last bodysurfed Hawaii.

People lay their boards on the grass of a nearby park and kick back with a free luau including salmon and veggie burgers, speeches on how to organize to protect coastal water quality, and a performance by a Hawaiian hula group.

I talk to Doug Hartley, a buffed-out black man in red swoopy sunglasses, a billcap, shorts, and a T-shirt. He has a cheerful, mellow demeanor and ready rap. Doug is the founder of Afrosurf Shack, a nonprofit youth outreach program and website.

"We're going to schools and setting up adventure trips. I've gone to Visitation Valley and James Wick Junior High Schools to reach out to kids. I bring my surfboard with me."

I ask him how long he's been surfing, and how many other African Americans he's seen in the waves. "I started surfing four years ago. I never saw too many black people surfing. There's lots of pollution from oil and chemical refineries in low-income neighborhoods like Richmond [a black and Latino community in the North Bay], and that tends to become the water issue people relate to.

"When I was sixteen I tried surfing on Long Island but I had a short board and it wasn't happening for me. I was 6 miles from the beach and didn't have any surfer friends in my black New York neighborhood, so instead of returning to the beach I got into skating. Later I went to Taos and became a skier and ski instructor but when I moved to California, I thought this is it. Seeing the surf culture, I decided I'd try again and moved to Pacifica to be by the water and here I am."

"Is Surfrider helping your program?" I wonder.

"When I first started my thing, Surfrider kind of helped out with information. 'Respect the Beach' was this 17-minute video they gave me but it was kind of boring. If I was a ninth grader, I'd go to sleep watching it. So I reedited it with more surfing, skating, and snowboarding scenes, kind of dropping the environmental messages in between. Of course, once

I've got the kids on the bus, that's a long ride and I play environmental movies on the bus video and what can they do?" he grins. "They've got to watch."

I'm back in New Jersey with Dery Bennett, the tall, gristled, 68-year-old director of the Littoral Society. Raised in Philadelphia, Dery first went to work at Woods Hole in the summer of 1951. "I wanted to do marine biology but soon realized most of their work was physical oceanography for the navy. We'd sit at the mouth of Cape Cod Canal and record the sound of vessels going in and out and then call the ships and ask them for their speed, direction, what kind of props they had. They were developing a sound catalog for identifying ships, I think."

On the Fourth of July in 1952, Dery decided if he couldn't beat them, he'd join the navy. He became an underwater demolition diver operating out of submarines. "You'd lock out of the sub, swim to the beaches, plant charges, then duck back into the water, escape into the lock as the sub sat on the bottom waiting for you. One time we hit the beach in Turkey, only no one had told the Turks we were coming so they started shooting at us."

After the navy he went into newspaper work in Philadelphia, did graduate work at Harvard, taught at a community college, and in 1968 became the society's first full-time staffer.

He invites me for a ride and seems to visibly brighten as we step outside his Sandy Hook office, and why not? There is a grand water view of New York Harbor, including a wooden pole fish trap about 100 yards out used to snare shad, herring, fluke, and menhaden. He tells me that there is a multi-million-dollar clam fishery in the harbor today, a sign of improving water quality, as is the increasing number of harbor seals. His own clamming rake and boots are in the back of his pickup.

We drive over the Shrewsbury River into Sea Bright. Thirty clam boats with big outboards sit by the dock next to the local clam house. Three rough-looking clammers are hanging out, playing with a dog. Inside the depuring house there are bags of quahogs sorted down, the small tasty ones just a little bigger than a silver dollar and worth about 25 cents each. They'll spend 48 hours in here being "depured," washed, and exposed to ultraviolet light to cleanse them. Back in the parking lot Dery takes his turn playing with the dog while I check out some of his bumper stickers: "Get Hooked on Fish Tagging," "Missing your cats? Check under my tires," "Invest in America. Buy a Congressman."

We drive through Highlands, an old rumrunner port, and on to the Twin Lighthouse built in 1863. From here you can see Coney Island, the World Trade Center, and downtown Manhattan along with 30 to 40 miles of Long Island's south shore.

More than a century ago young striped bass were caught here in the Navesink River and shipped by train to San Francisco, hardy fingerlings traveling across country in milk cans. Today, the descendants of these pioneer stripers are the indicator species for the health of San Francisco Bay.

On the East Coast, stripers are heralded as a conservation success story. Fished almost to extinction between the 1970s and 1980s, cooperative efforts by federal and state agencies, commercial and recreational fishermen, as well as ongoing environmental cleanups of their habitats in the Hudson River and Chesapeake Bay, have resulted in a dramatic comeback of these feisty predator fish that can grow as large as bulldogs.

We drive over to Belford, New Jersey, past wetland rivers full of geese and egrets and the clacking sound of clapper rails. This working-class fisherman's holdout on the gentrifying Jersey shore has a sordid reputation. Fish cops have told me Belford is the world capital of undersized lobsters. There are wooden reels for net drying behind the fish co-op and Pirate Cove Restaurant at the end of a shell and gravel parking lot. A few guys are offloading barrels of menhaden, an "industrial fish" used for oil and chicken feed. Across a narrow, black-water channel a younger man with scraggly dark hair, stained jeans, and a T-shirt is taking live horseshoe crabs out of his Boston whaler and cutting them in half with what looks like a giant paper cutter. As he pulls the blade handle down it sounds like he's slicing through wet cardboard. He tosses the tails in the water and dumps the crab halves in a box. Used for eel bait, they sell for a dollar an animal.

The Highlands Historical Society has asked Dery to speak this evening. He tells me this as we sit backed up in traffic behind Range Rovers, BMW roadsters, and other upscale cars headed to the new ferry terminal connecting northern Jersey and lower Manhattan. A man passes us in a Lexus, talking on a cell phone. Dery picks up a heavy black phone receiver he keeps on the dash of his truck and pretends to talk into it.

"I'm all for more ferry service. I just worry about the Yuppies it attracts," he tells me. Driving back through flood-prone downtown Highlands, we see parents and kids with their dogs lined up in front of the fire station where they're giving free rabies shots.

Later I go to the community center for the historical society meeting. It's a cheaply built wooden building with a nice playground next to the beach. There is a children's mural of dolphins and fish and a big picture window looking out on the water. Dery arrives and speaks briefly with the society chairwoman. There are some 25 people gathered this evening, mostly seniors. Dery talks to groups like this a couple of times a week. In the age of the Internet and global mass media, he believes personal contact is still the most effective form of organizing.

"I had one of your bumper stickers up in my office," a legal services lawyer who works in Newark tells him. It said, 'Have you hugged your estuary today?' One of my clients read it and when I asked her if she had, she hugged me. I thought this was cute and told my boss, but from his reaction I could see he didn't know what an estuary was either."

"Seventy percent of the earth is covered by ocean, the other 30 percent is optioned by developers," Dery tells the group before going on to explain that "littoral" is Latin for the shoreline. He defines littoral as from the head of tides up rivers to the edge of the continental shelf. Outside the sky is turning pink above the New York bight as a small sailboat beats against the wind.

"Wherever you build seawalls and other structures, the ocean curls around behind and begins to erode them," he explains. "If you want to make a beach disappear, build jetties and seawalls. Also, at least on the Atlantic coast, you have both sea-level rise and land subsidence going on at the same time. Geologists say it doesn't matter which, your feet are still going to get wet."

"But things are getting cleaner aren't they?" someone asks.

"Some places yes, some no."

"So what's your feeling about the future?"

"If you're asking if I'm optimistic or pessimistic, I'd have to say it depends on what day and what time of day." He then starts talking about the local things that make him proud.

"In New Jersey crabbing is the number-one ocean sport, more than fishing and it's all about blue crabs. You can go after dungeness crabs or others crabs elsewhere, but why bother? Also, let me tell you about the American eel and the European eel. They're examples of amazing systems that animals develop to reproduce.

"Eels are born in the Sargasso sea. The elvers, or baby eels, they're also called glass eels, start up the East Coast adrift in the Gulf Stream. The elvers of the American species begin to actively swim into rivers and bays of North America. The Europeans drift on past Greenland. Now, a few years ago these elvers were worth up to $300 a pound because of Japanese demand, where they raise them in ponds till they're a few inches long (he holds two fingers slightly apart) and then eat them as a delicacy." The woman next to me scrunches her face in disgust.

"Males stay in the saltier water of the estuaries while the females go upstream. They can climb dams or swim across damp ground to get to freshwater ponds. If you want to come tomorrow, we'll put on waders and go to a dam I know and I'll show you little gray spots on the dam face and if you look closely you'll see it's groups of 30 or so tiny eels."

No one raises a hand to volunteer for his field trip.

"After three to seven years they grow big, a few feet long, 4 to 8 pounds, and then the females, who are now much bigger than the males, head back downstream. There they get fished with horseshoe crab meat. The survivors meet males in the estuaries and swim to the Sargasso Sea where they spawn and die. Meanwhile, the Europeans stall their own development for as much as a year in the Gulf Stream as they drift past Greenland. They won't mature till they get to their European rivers."

He gets some follow-up questions, mainly about how to control the glass eel fishery.

"We've gotten some controls on taking horseshoe crabs, and I think we can do the same with this," he tells them.

State agencies have in fact begun cracking down on the glass-eel fishery. It also helps that with Japan's recent economic recession the going rate for elvers crashed to around $10 a pound, leaving little incentive for illegal poaching.

Dery talks about how he used to fish for big eels and how people should eat more local food—steamers and eels and blue crab. Today people would rather go to the Jersey Shore and eat snapper from the Gulf of Mexico or shrimp farmed in Asia and Latin America. "It's too easy to eat exotic and forget to protect your own local abundance," he warns.

One younger woman who recently moved back to Highlands describes how, while growing up, her family would catch eels on the shore and take them home, where they'd nail them to a tree in the backyard and strip their skin off and boil them, watching the dead eels curl in the bubbling water. The way she recounts this scene it sounds like she's describing her baby's first steps. Still, as organizers like Dery Bennett know, it is out of just such unique and parochial traditions that ocean protection movements are born.

"I believe all the ocean issues will get dealt with when people demand we take action," says Curt Weldon, the Republican representative from Pennsylvania. "My district's not on the ocean but my people go to the ocean to enjoy themselves," he says. "I like to boat and fish, and I see the ocean as a glue that can bring people together."

In 2000, Weldon, along with Jim Saxton of New Jersey, Democrat Sam Farr of California, and a few others, helped form a new oceans caucus in the House of Representatives. Weldon hopes to eventually see an oceans committee in the House to replace the Merchant Marine and Fisheries Committee his party abolished in 1995. "Protecting the oceans is too important to be seen as a partisan issue," he argues. "No party can own it.

Fishermen, oil interests, environmentalists, we all need to protect the oceans. We all need to start thinking more about biology and ocean governance and security and pollution."

Thirty-five million people a year now visit the Jersey Shore. An equal number go to New York's Jones Beach. Even after a huge decline due to concerns over water quality, the beaches of Los Angeles still attract some 20 million visitors. There are 20 million annual visitors to our national seashores, and even more to our marine sanctuaries. About 12 million Americans are into recreational saltwater fishing; 10 million like to whale watch; 8.5 million Americans are certified scuba divers; and at least an equal number surf, windsurf, bodyboard, or surf kayak.

We are eating more seafood than ever before and moving to the beach in record numbers. We love the ocean, we use the ocean, but we don't think enough about the ocean. We've mapped only 10 percent of the ocean with the same resolution as we have mapped 100 percent of the moon. If you're into lifeless places, you should visit the moon, travel to Mars, or check out Washington, D.C. If you want to see life on its own phantasmagorical terms, go to sea. If today we are loving our oceans to death through ignorance and short-term greed, we can stop doing that. After all, the hope of any new frontier is that we do not have to repeat the mistakes of our past. Of course, the ocean itself is monumentally indifferent to all of our human hopes and desires.

During a big California storm surge in the fall of 1999, a honeymoon couple went to Lovers Point in Pacific Grove near Monterey. The sea was as wild as their passion that day. Jennifer Stookesberry, 23 years old, stood with her back to the raging surf posing for her new husband, 28-year-old Eamon Stookesberry. Suddenly, a huge sleeper wave broke, dragging her backward into the sea. Eamon dove off the cliff after her, grabbed her hand, but the sea pulled her away. He called to her to stay afloat, not to panic as the currents took her farther out. "She got really far away and I heard her scream for help. Then there were two big waves and that was the last time I saw her." The Coast Guard rescued Eamon and pulled Jennifer out of the surf. Her eyes were open but dark. They were unable to revive her. He held her hand again at the hospital, in a special way they had, to show off their matched rings, and quietly said his goodbye. A day later he met with reporters. "If you people have anybody special, don't sweat the small stuff, make it work," he told them. "You get one chance. She was that one for me. She was my angel, my kitten."

"I really don't know why it is that all of us are so committed to the sea, except I think . . . it's because we all came from the sea," said John Kennedy at the 1962 America's Cup races. He was a sailor and former navy man who had survived war and his PT boat's ramming and sinking at sea, only to be felled by an assassin's bullet. His son, John, Jr., died in a plane crash at sea and was buried at sea with his wife and sister-in-law.

In music, art, and literature, tragic, wild, and inspirational, the Blue Frontier continues to ripple through the American soul. From sea chanteys to the Hawaiian hula to Jimmy Buffett singing of how "You can't reason with hurricane season," the oceans have offered their syncopated inspiration to American musicians of every age, race, and disposition.

I remember on my birthday a few years back, after an overnight anchorage and morning dive with a curious seal at the Coronado Islands 13 miles off San Diego, my friends and I were headed back to the bay. Crosby, Stills, and Nash were on the sound system, and there were Pacific white-sided dolphins riding our bow wave. My friend Jon Christensen was at the wheel of the 37-foot sloop. I was leaning back against the bulkhead with a Corona beer in one hand, while Scott Fielder was packing away his dive gear. The song "To the Last Whale" began playing, an arrangement that includes recorded humpback whale song.

"All we need now is a whale," I grinned at Scott. "There's one," he nodded casually. Just off to our starboard side rolled the crusty back followed by the broad paddle tail of a California gray whale. Just another frontier moment.

Of course, I number myself among the tens of millions of Americans who have found their way back to the sea at least in part through visual media including film and television. From *Gidget* to *The Endless Summer*, *Jacques Cousteau*, *Sea Hunt*, and *Flipper*, to *Jaws*, *Waterworld*, and *Baywatch*, from the sublime to the ridiculous, the ocean remains a popular theme for popular culture.

Today, with improved underwater film techniques and lighting, dive technology and communications, the chances to appreciate the wild beauty of the Blue Frontier without actually getting wet have also expanded exponentially. *National Geographic* and other television documentaries have begun to include point-of-view wildlife images using detachable "critter cams" carried on the backs of dolphins, seals, white sharks, and blue whales. Real-time images from Aquarius, the deep ocean, and other frontier realms are also being brought to the classroom and online via educational projects like JASON and collaborative efforts between broadcasters and scientists. Giant IMAX screens around the country offer large-as-life ocean images, including 3-D panoramics of marine wonders. Many poets have also found inspiration in the living sea, from "She sells sea shells by the seashore," to T. S. Eliot's *The Dry Salvages:* "The sea has many voices. Many gods, and many voices."

In painting, Winslow Homer was perhaps America's greatest maritime artist. He showed the Blue Frontier in its fathomless indifference to those who seek to live and work upon its waters or, conversely, its hypnotic attraction for those who have survived the ravages of civilization (like the American Civil War he also illustrated) and now seek its solace.

From the sail-happy youth of *Breezing Up* (1876) to the edgy wonder of a fisherman in his dory in *Fog Warning* (1885), to the simple interface of land and sea in *West Point, Prout's Neck* (1900), Homer offered the public a vision of redemption through saltwater immersion. Whether in the stormy challenge of *The Lifeline* (1883) or the easy labors of *Key West Hauling Anchor* (1903), his imagery won the nation by rejecting the gauzy romanticism of the Hudson River school and other self-conscious stylists. Much of today's maritime art by contrast tends toward the kitschy—smiling dolphins, leaping whales, and come-hither mermaids swimming through reefs of clown fish.

California artist Wyland has opted to instead go for the monumental, with his building-sized whale murals gracing cities across America and inspiring many a reflective thought among commuters stuck in traffic from Boston to San Diego, Orlando to Seattle.

Alaskan illustrator Ray Troll combines humor and subtle anger in his depictions of the Blue Frontier's sealife under siege. Faced with a "bass ackwards" world, he forces us to confront our genetic connections in images like *Spawn Till You Die,* showing a human skull above crossed salmon, with naked people and fish floating about the perimeter. A gentler prod to our conscience comes in the daily comics section of more than 150 newspapers in the form of *Sherman's Lagoon.* Jim Toomey's Sherman is a shark, sharper of tooth than of mind. He, along with Fillmore the sea turtle, Hawthorne the hermit crab, and other characters, remind us that we are not the only critters on this blue planet and that the food chain can nibble up as well as down. Of course, almost every marine scientist you meet pines for the day Gary Larson might return to draw more *Far Side* panels with their anthropomorphic reversals of life and nature, including the one of biologists wondering what the dolphins' vocalizations might mean as one of the cetaceans in the tank vocalizes, "Habla Español?"

There is also the utilitarian art of surfboard shapers like Skip Frye, boat designers like Hobart "Hobie Cat" Alter, and underwater photographers David Dubuoilet, Norbert Wu, Flip Nicklin, and Al Giddings.

Richard Henry Dana's *Two Years before the Mast* and Herman Melville's *Moby Dick* gave nineteenth-century America a realistic sense of the sea, as did authors Jack London, John Steinbeck, Rachel Carson, and others in the twentieth century. Among contemporary writers, John

McPhee, Daniel Duane, Robert Stone, and Sebastian Junger impress me as having the stoke, although this kind of listing can be as idiosyncratic as whoever chooses to compose it. Perhaps some of the finest ocean expressionism I have seen is in children's murals, drawings, and artworks at aquariums and day-care centers, in classrooms, and at marine fairs and science centers. Of course, kids have an advantage, being closer to the source. Fear of water, it turns out, is an acquired, not a natural, state of being.

It's hard to find an ending in writing of a frontier without end, of a vast mixing place for water, air, earth, life, and grace. I've been lucky in my explorations: I have survived near-drowning in big surf off Oahu's north shore, been humbled by wild waves and salmon in Alaska, found peace in the tranquil waters of the U.S. Virgin Islands, rode a whaleshark, shipwrecked a sailboat in the Sea of Cortez south of San Diego, and met unforgettable watermen and -women whenever I've ventured on or near the Blue Frontier.

Right now I am standing on a cliff at Point Lobos State Park, just south of Monterey. From here I can look to my left and see several rocky offshore islands covered with hundreds of lazing sea lions who have waddled up to the highest crags, 50 feet above the water. The surging sea breaks turquoise and white below them. In front of me is a horseshoe-shaped cove under wind-sculpted cliffs topped by thick stands of dark green cypress and pine. There is a crescent beach where several gray and black spotted harbor seals have hauled up on the coarse sand. A raft of half a dozen sea otters is lolling in the kelp, rolling themselves in the green translucent plants like so many servings of otter sushi. Pelicans are making kamikaze dives on a shoal of sardines. Cormorants and gulls are flying about like busy rush-hour commuters. A great blue heron stands on a floating piece of plywood, looking down into the water, poised like the predator that he is. He strikes below the surface with his rapier bill and tumbles in after it, sinks, and resurfaces, flapping back onto his float in an undignified ruffle of feathers. The wildlife here is as abundant and easy to spot as my own species back in town and equally disinterested in my observing them.

The only place I know with a richer variety of life is below the water's surface. I lean into the sea wind and recall last night's video feed from the Monterey Bay Aquarium's deep-diving robot sub. It was 600 feet down in the cold black canyon's midwater range when a white shark, wide as a jet fighter's fuel tank and sleek as death, swam through the ROV's spotlight beam. A visceral shot of adrenaline ran through my body, reminding me of the Edward Abbey quote, "If there's not something bigger and meaner than you are out there, it's not really a wilderness." Our waters remain rich in potential adrenaline jolts and revelations, in waves that call to be ridden, winds made to snap a sail, shells newly tossed upon the shore, sunsets not to be believed.

Our oceans remain full of strange wonders and grand experiences that will thrill generations yet unborn. Despite all the problems and challenges we face fighting for America's living seas, that is still enough to give one hope. After all, it is not every great nation, forged by its earliest frontier experiences, that gets a second chance.

Notes and References

Introduction Thrashed

p. 2, *Exclusive Economic Zone (or EEZ)*: Presidential Proclamation 5030, March 10, 1983. Only 532 words, the proclamation established the world's largest EEZ, an area of some 4 billion acres.

p. 3, *icy crust of the Jupiter moon Europa*: For an in-depth look at the possibility of life on Europa and the theories that support it, see "Is There Life Elsewhere in the Universe?" by Jill C. Tarter and Christopher F. Chyba, *Scientific American*, December 1999.

p. 3, *Among these is* Pyrococcus: For a good telling of the discovery of *Pyrococcus*, see William Broad, *The Universe Below* (New York: Simon & Schuster, 1997), 278–281.

p. 4, *according to the United Nations Food and Agriculture Organization*: UN FAO, *The State of World Fisheries and Aquaculture* (Rome, Italy: UN FAO, 1995), 8.

p. 5, *or you can be eaten by great beasts*: Interview with Ken Kelton, reported in "Great White Comeback," *Men's Journal*, June/July (1996): 133.

p. 6, *after 107 years of operation*: U.S. Congress, *Final Report on the Activities of the Merchant Marine and Fisheries Committee* (Washington, D.C.: Government Printing Office, 1995).

p. 6, *upcoming New Hampshire presidential primary*: "When $5 Million Doesn't Add Up," *New York Times*, September 12, 1999.

p. 7, *pass the Oceans Act*: Information from Representatives Jim Saxton and Sam Farr and Washington congressional, NOAA, and U.S. Navy staffers.

p. 8, *may very well be saving ourselves*: As quoted in Douglas Gantenbein, "Making Room for Salmon," *Sierra Magazine*, July/August (1999): 19. The same quote ran in several other publications.

Chapter 1 Fool's Gold

p. 12, *recovering several tons of nodules*: David Helvarg, "Race to Control the Sea Floor," *In These Times*, December 14, 1977.

p. 12, *depths in the mid-Pacific*: Elisabeth Mann Borgese, ed., *Ocean Frontiers* (New York: Abrams, 1992), 9–10.

p. 12, *space exploration was then receiving*: Elisabeth Mann Borgese, ed., *Ocean Frontiers* (New York: Abrams, 1992), 48, and various interviews.

p. 13, *adopted by the UN in 1970*: Clyde Sanger, *Ordering the Oceans* (London: Zed Books, 1986), 18–20.

p. 14, *before the 1972 elections*: Sherry Sontag and Christopher Drew, *Blind Man's Bluff* (New York: Public Affairs, 1998), 75–85.

p. 14, *added weight to the charge*: Nathan Miller, *Stealing from America* (New York: Paragon House, 1992), 330.

p. 14, *a manganese mining ship*: Clyde W. Burleson, *The Jennifer Project* (College Station: Texas A&M University Press, 1997), 57.

p. 16, *manganese nodules for publicity purposes*: Clyde W. Burleson, *The Jennifer Project* (College Station: Texas A&M University Press, 1997), 76, 82.

p. 16, *noted on receiving the report*: Committee On Commerce National Ocean Policy Study, *The Economic Value of Ocean Resources to the United States* (Washington, D.C.: Government Printing Office, December 1974), 20–22, and letter of transmittal.

p. 17, *one light in the Pacific never went on*: Interview with Michael Molitor, now director of the University of California Climate Change Program, November 5, 1999.

p. 17, *ambassador to the Law of the Seas convention*: Interview with Elliot Richardson, July 1, 1999. Richardson died of a cerebral hemorrhage at age 79 on December 31, 1999.

p. 17, *state of complete chaos*: Interviews with Robert Friedheim in 1995 and November 18, 1999.

p. 18, *before the Lord returns:* Philip Shabecoff, *A Fierce Green Fire* (New York: Farrar, Straus, 1993), 208.

p. 19, *strategic minerals for future defense*: William J. Broad, *The Universe Below* (New York: Simon & Schuster, 1997), 262.

p. 19, *newly discovered life-forms*: Richard Charter, *A Citizen's Guide to Ocean Stripmining* (San Francisco: Gorda Ridge Project, 1984).

p. 19, *not going to be his priority*: Interview with a Helms Senate Foreign Relations staffer, November 23, 1999.

p. 20, *nickel market is going to recover*: Michael D. Lemonick with Andrea Dorfman, Irene M. Kunii, Alice Park, and Tala Skari, "Mysteries of the Deep," *Time* cover story, August 14, 1995.

p. 20, *according to a number of studies*: The earliest study to raise these concerns was the September 1981 "Environmental Impact Statement on Deep Seabed

Mining," published by NOAA's (since dismantled) Office of Minerals and Energy.

p. 21, *found that out years ago*: Interview with Abraham Piianaia at Hawaii Maritime Center, spring 1990.

p. 22, *with many people in marine science*: Interview with Bill Duoros, September 18, 1999.

p. 22, *it was all classified*: Interview with Skip Theberge, April 9, 1999.

p. 23, *as we broke free*: Interviews with Gary Greene in 1996 and on November 23, 1999.

p. 23, *didn't make it back up*: Interviews with Rich Slater in 1996 and 1999.

p. 24, *creating itself at these chimneys*: Interviews with Don Walsh in 1996 and on December 4, 1999.

p. 24, *in the years since*: National Research Council, *Undersea Vehicles and National Needs* (Washington, D.C.: National Academy Press), 7.

p. 25, *along the bay's crescent shore*: Chris Grech has since been promoted to assistant director of marine operations at MBARI. The marine institutes have organized as the Monterey Bay Crescent Ocean Research Consortium.

p. 25, *getting down there intrigued him:* Interview with Julie Packard, 1996.

p. 26, *right tools for the job*: Interviews with Don Walsh in 1996 and on December 4, 1999.

p. 27, *it kept the boss happy*: Interview with Chris Grech, December 9, 1996.

p. 28, *and see what we get*: Interview with David Newman, November 23, 1999.

p. 28, *where 80 percent of the planet's life-forms exist*: *Turning to the Sea: America's Ocean Future.* Report from Vice President Al Gore's White House Office (September 2, 1999), 22. Copies available from U.S. Department of Commerce/NOAA.

p. 29, *You're a public institution*: Interviews with Zeke Grader, April 14 and September 3, 1999.

p. 29, *that year's congressional elections*: Figures are from the Center for Responsive Politics (www.opensecrets.org), which breaks down the tidal wave of money flooding Washington using data from the Federal Election Commission, Lobbying Disclosure Act of 1995, and other public records.

p. 29, *the fishermen who depend on them*: David Attaway, ed., "Aquacultural Endocrinology and Molecular Genetics," *Sea Grant—The Sea's New Harvest: A Report on Marine Biotechnology in the National Sea Grant College Program* (Washington, D.C.: NOAA), December 31, 1996, 3.

p. 30, *frontier prospectors as fool's gold:* Interviews with Dr. John DeLaney, December 20, 1999, and March 8, 2000, and PBS *Nova* special, "Volcanoes of the Deep," produced by Susan K. Lewis, 1999.

Chapter 2 **From Sea to Shining Sea**

p. 33, *America's oceanic immigrants*: David Sean Paludeine, ed., *Land of the Free* (New York: Gramercy Books, 1998), 33–34.

p. 34, *and hauling them back up*: For the definitive history of cod, plus some good recipes, read Mark Kurlansky, *Cod* (New York: Penguin Books, 1997).

p. 34, *in the local timber trade*: Richard Hofstadter and Michael Wallace, *American Violence* (New York: Vintage, 1971), 110–111.

p. 35, *in Boston in 1768*: Pauline Maier, *From Resistance to Revolution* (New York: Vintage, 1972), 6–7.

p. 35, *winter of both the poles*: Charles A. and Mary R. Beard, *The Beards' New Basic History of the United States* (New York: Doubleday, 1960), 51.

p. 37, *hundreds of men were lost*: The author gathered information on whaling while preparing an audio tour for the Hawaii Maritime Center in 1990. Additional information can be found at the New Bedford Whaling Museum.

p. 38, *assigned to their growth*: From the display "Pathfinder of the Seas," U.S. Navy Museum, Washington Navy Yard, Washington, D.C. See also Naval Oceanographic Office history page online at www.navo.navy.mil. For a different perspective on civilian versus naval hydrographic history (and the reef report), visit the NOAA library history page at www.NOAA.gov.

p. 38, *grain bound for northern Europe*: From a display in the San Francisco Maritime Museum.

p. 39, *arrived as railroad tourists in the 1870s*: According to Shelley Lauzon, senior news officer (and occasional historian) for the Woods Hole Oceanographic Institution.

p. 39, *cold and treacherous Pacific*: For more on Sutro Baths and Coney Island, see Lena Lencek and Gideon Bosker, *The Beach* (New York: Viking, 1998), 163–171.

p. 41, *America went to war*: For a fuller description of the navy's growth between the Civil War and World War I, see chapter 7, "Not Merely a Navy for Defense," in Kenneth J. Hagan, *This People's Navy* (New York: Free Press, 1991).

p. 43, *death toll exceeds 25,000*: David McCullough, *The Path Between the Seas* (New York, Simon & Schuster, 1977), 610.

p. 43, *24-hour radio watches*: From research the author conducted for the audio tour "Titanic: The Exhibition," written with Sally Rudich. Also, author's reporting for "Lessons from the Deep," *George* magazine, December 1997.

p. 44, *between Key West, Florida, and Havana*: Benjamin W. Labaree et al., *America and the Sea: A Maritime History* (Mystic, Conn.: Mystic Seaport, 1998), 524–530.

p. 45, *off its elevated rail bridge*: PBS documentary on Flagler (1998) and various history books.

p. 46, *Marine Biological Laboratory already existed*: Elisabeth Mann Borgese, ed., *Ocean Frontiers* (New York: Abrams, 1992), 63–65.

p. 46, *hit the waterfront hard*: Luc Cuyvers, *Sea Power* (Annapolis, Md.: Naval Institute Press, 1993), 233–235; and Donald W. Cox, *Explorers of the Deep* (Maplewood, N.J.: Hammond, 1968), 33–36.

p. 47, *and improved pay*: Richard O. Boyer and Herbert M. Morals, *Labor's Untold Story* (Pittsburgh, Pa.: United Electrical Radio and Machine Workers of America, 1997), 282–289.

p. 47, *Kennedy pledged*: Labaree et al., *America and the Sea*, 541.

Chapter 3 Admirals and Oceanographers

p. 50, *which was 118 feet wide*: Interviews conducted for Pearl Harbor tour produced by Antenna Audio in 1991.

p. 53, *came into their own*: Interviews with Walter Munk in 1984 and on October 14, 1999.

p. 54,. *research for the navy*: Judith and Neil Morgan, *Roger* (San Diego, Calif.: Scripps Institution of Oceanography, 1996), 33.

p. 54, *University of Texas at Austin*: National Research Council, *Oceanography in the Next Decade* (Washington, D.C.: National Academy Press, 1992), 32–33, and interviews.

p. 54, *any of the ships decontaminated*: Interview with Roger Revelle for the author's "Three Giants Who Were There," *San Diego* magazine, August 1982.

p. 55, *in February 1949*: John E. Pfeiffer, "The Office of Naval Research," *Scientific American*, February 1949.

p. 55, *and later succeeded him*: Interview with Art Maxwell, November 19, 1999.

p. 55, *go out to sea together*: Interviews with Gary Weir, May 10 and May 13, 1999.

p. 56, *marine science into the 1970s*: Edward Wenk, Jr., *The Politics of the Ocean* (Seattle, Wash.: University of Washington Press, 1972), 13.

p. 56, *also good for the navy*: Interview with Art Maxwell, November 19, 1999.

p. 56, *the "revolt of the admirals"*: The revolt of the admirals is recounted in exhausting detail in Jeffrey G. Barlow, *The Revolt of the Admirals* (Washington, D.C.: Naval Historical Center, 1994).

p. 56, *wants to see another one*: Interview with Roger Revelle for the author's "Three Giants Who Were There," *San Diego* magazine, August 1982.

p. 57, *advanced guidance systems*: Borgese, *Ocean Frontiers* (1992), 87–92, and Project Nobska files and interviews at Woods Hole Oceanographic Institute.

p. 57, *senior archivist at Scripps*: Interview with Deborah Day, October 14, 1999.

p. 58, *such heavy security*: Interviews with Walter Munk in 1984 and on October 14, 1999.

p. 58, *using it for years*: Interviews with Gary Weir, May 10 and May 13, 1999.

p. 59, *the ship itself*: Laurence Gonzales, "Ballard Surfacing," *National Geographic Adventure,* Spring 1999.

p. 59, *next to one of the corrals*: This was part of a tour the author was given of the Space and Naval Warfare Systems Center, San Diego, on October 19, 1999.

p. 60, *in shallow waters*: Tony Perry, "Navy Sonar System Draws Activists' Fire," *Los Angeles Times,* October 14, 1999, and interviews with two navy sources.

p. 60, *denied any connection*: Rick Weiss, "Whales Became Stranded After U.S. Naval Exercises," *Washington Post,* March 22, 2000.

p. 60, *as a navy technical report*: Stephen Leatherwood et al., "The Whales, Dolphins, and Porpoises of the Eastern North Pacific: A Guide to Their Identification in the Water" (San Diego, Calif.: Naval Underseas Center TP 282, 1972).

p. 61, *"swimmer nullification"*: From a review of Greenwood's testimony; copies were provided to a number of journalists, including Steve Chapple and Robert E. Kesslen, who wrote the story up for *Newsday,* April 11, 1976. Also, author's interview with Greenwood from 1985.

p. 62, *talking to each other*: The Ken Woodal interview and other original research referred to in this chapter were for several of the author's articles, including "Marine Mammals Serving the Navy," *Pacific News Service,* June 1985.

p. 62, *for revealing classified information*: Interviews with Rick Trout in 1988, 1999, and 2000; and Jane Fritsch. "Company Fires Animal Trainer Who Criticized Navy Program," *Los Angeles Times,* November 3, 1988; and "Dolphin Trainer May Be Prosecuted," Associated Press, reprinted in *San Francisco Chronicle,* November 11, 1988.

p. 62, *to get away from it*: Dianne Dumanoski, "Former Trainer Objects to Military Use of Marine Animals," *Boston Globe,* November 9, 1990; and interview with David Reames, November 1990.

p. 63, *amphibious landing zones*: Department of the Navy, *Vision, Presence, Power: A Program Guide to the U.S. Navy* (Washington, D.C.: Government Printing Office, 1999). 102.

p. 63, *That's our goal*: Interview with Rear Admiral Paul Gaffney, January 6, 2000.

Chapter 4 Quarrel on the Littoral

p. 67, *and fair winds*: "Oceanography: The Making of a Science Dinner," dinner program, March 8, 2000.

p. 67, *over the Golan Heights*: From the CIA website (www.cia.gov) and interview with Linda Zall, March 8, 2000.

p. 68, *how you make your case*: Interview with Representative Curt Weldon, March 30, 2000. First brief quote is from his article "Ocean Action 1999: Some—But Still a Long Way to Go!" *Sea Technology,* January 2000.

p. 68, *California and Massachusetts*: MEDEA, *Scientific Utility of Naval Environmental Data* (McLean, Va.: MEDEA, 1995); and MEDEA, *Ocean Dumping of Chemical Munitions: Environmental Effects in Arctic Seas* (McLean, Va.: MEDEA, 1997). The author found little federal desire to fund new research on ocean dumps while producing a segment for a syndicated TV show. Rear Admiral Andrew Granuzzo confirmed this lack of interest.

p. 69, *plaque in his honor*: Gordon J. Peterson and David E. Werner, "Under the Sea, at the Top of the World," *Seapower,* July 1999; and Woods Hole Public Relations Office on Al Vine.

p. 69, *missions will go unfulfilled*: Peterson and Werner, "Under the Sea."

p. 69, *director of environmental protection*: Interview with Rear Admiral Granuzzo, January 7, 2000.

p. 69, *environmental data collection*: Jeffrey T. Richelson, "Scientists in Black," *Scientific American,* February 1998. The presidential directive came out in 1993.

p. 70, *for CO_2 reduction*: Interview with Charlie Kennel, October 14, 1999.

p. 70, *than we do today*: Interview with Bob Gagosian, March 8, 2000.

p. 70, *and of cods*: Interview with Deborah Day, October 14, 1999.

p. 70, *something that might matter*: Interview with Charlie Kennel, October 14, 1999.

p. 71, *to carry out its mission*: Highlights of the Department of the Navy FY 2000 Budget; and Bradley Graham, "Pentagon Warns Against Cutting Attack Sub Fleet," *Washington Post,* January 4, 2000.

p. 72, *lost his lunch*: Visit to the USS *John C. Stennis,* October 15–17, 1999.

p. 72, *and the UN*: From "Welcome Aboard" flyer, USS *Stennis,* and presentations by Lieutenant David Oates, public affairs officer.

p. 74, *navy insists they should be*: *Navy Aircraft Carriers: Cost-Effectiveness of Conventionally and Nuclear-Powered Carriers* (Washington, D.C.: General Accounting Office/NSIAD-98-1, August 1998).

p. 74, *new and exciting ways*: Admiral Jay Johnson's address at the Naval Institute's annual meeting, April 21, 1999.

p. 78, *until his retirement*: Three people who served under Rear Admiral George Davis, appointed director in December 1992, recounted to the author how morale in the oceanographic office sank during his command tenure.

p. 79, *Hobart [Australia] last year*: Interview with Lieutenant Commander Paul Kratochwill, October 16. 1999.

p. 79, *Naples, Cannes, Greece*: Interview with Rear Admiral Granuzzo, January 7, 2000.

p. 80, *Thirty-second Street Naval Station*: National Oceanographic and Atmospheric Administration National Status and Trends Program, "Sediment Toxicity in U.S. Coastal Waters," April 1998.

p. 80, *20mm and 50mm rounds*: Numerous clippings, including Terry Rodgers, "Navy Says Ordinance Ends Plan for Sand," *San Diego Union Tribune*, December 16, 1997.

p. 80, *by a navy bomb*: Numerous clippings, including Roberto Suro, "President Intervenes on Vieques," *Washington Post*, November 17, 1999.

p. 81, *center of the island*: Interview with Rear Admiral Granuzzo, January 7, 2000.

p. 81, *"excellence in all warfare areas"*: Reprinted by various Puerto Rican news organizations. After the ads were first publicized by protestors, the navy removed them from their Vieques website.

p. 81, *chemicals in U.S. waters*: Mathew L. Wald, "Royal Caribbean Admits It Dumped Oil and Hazardous Chemicals," *New York Times*, July 22, 1999.

Chapter 5 Oil and Water

p. 84, *takes place in the gulf*: Percentages from Public Affairs Office, Mineral Management Service.

p. 84, *back in 1947*: Curtis Rist, "Why We'll Never Run Out of Oil," *Discover* magazine, June 1999.

p. 84, *12 stories above the water*: Visit to Pompano and Skipjack platforms, December 3–4, 1999.

p. 87, *an end to the practice*: Interviews with oil rig personnel, charter boat operator, former Louisiana state legislator, and National Marine Fisheries Service agent.

p. 87, *2600 feet of water*: "The Genesis Development," *Go Gulf* magazine, July/August 1999.

p. 90, *San Jose Mercury News in 1901*: Cited in Robert Jay Wilder, *Listening to the Sea* (Pittsburgh, Pa.: University of Pittsburgh Press, 1998), 33.

p. 90, *natural and unproblematic*: Robert Gramling, *Oil on the Edge* (Albany, N.Y.: State University of New York Press, 1996), 40.

p. 91, *to come in there and drill*: Wilder, *Listening to the Sea*, 52.

p. 92, *booms in American history*: Quoted from *The Promise and the Reward: 50 Years Offshore*, a 33-minute video produced by the National Ocean Industries Association, 1997.

p. 93, *movie theater at the time*: Cited in Daniel Yergin, *The Prize* (New York: Simon & Schuster, 1991), 754.

p. 93, *a marvelous business*: Bush interview from the video "The Promise and the Reward."

p. 93, *competition with customs tariffs*: Interview with Tom Kitsos, deputy director, Mineral Management Service, June 23, 1999, and Mineral Management Service Public Affairs, District of Columbia and Gulf of Mexico, December 21, 1999.

p. 93, *a bright future for all*: *Thunder Bay* (1953), Universal Pictures; directed by Anthony Mann and starring James Stewart, Joanne Dru, Gilbert Roland, and Dan Duryea.

p. 94, *skin, eyelids, and hair*: For a good description of Santa Barbara's response to the spill, see Marc Mowrey and Tim Redmon, *Not In Our Backyard* (New York: William Morrow, 1993), 15–19.

p. 95, *tender mercies of OPEC*: Don Hodel's speech taped by author and coproducer Steve Talbot for *Troubled Water*, PBS documentary that aired on California stations, Spring 1986.

p. 95, *Prince William Sound*: John McPhee, *Coming into the Country* (New York: Farrar, Straus, 1976), 127.

p. 96, *just not able to let it go*: Two good reports on Prince Williams Sound 10 years later are John Mitchell, "In the Wake of the Spill," *National Geographic*, March 1999; and Charles Siebert, "After the Spill," *Men's Journal*, April 1999.

p. 96, *an underwater pipeline*: "Chevron Pipeline Spills Oil on Grand Isle Barrier Island," *Baton Rouge Advocate*, November 25, 1999; radio reports; "Gas Platform Evacuated after Blowout in Gulf," *Times-Morning Advocate*, December 4, 1999; "Pipeline Break Spreads Oil in Gulf of Mexico," Associated Press report, *Washington Post*, January 23, 2000; "Only a Sheen Is Left from Gulf Oil Slick," Associated Press, January 24, 2000.

p. 97, *this tough new liability law*: Wilder, *Listening to the Sea*, 168–169.

p. 97, *ask Congress for relief*: Quoted from "OPA 90 Revisited," *Work Boat Magazine*, December 1999.

p. 97, *safeguard the environment*: Joan Biskupic, "High Court Overturns State Law on Oil Tankers," *Washington Post*, March 7, 2000.

p. 98, *of business and property*: Charles Fried, *Order and Law: Arguing the Reagan Revolution—A Firsthand Account* (New York: Simon & Schuster, 1991), 183.

p. 98, *a herring-baited lobster pot*: For more on takings battles and other environmental law conflicts, see the chapter "Up Against the Law" in David Helvarg, *The War Against the Greens* (San Francisco, Calif.: Sierra Club Books, 1994; paper edition, 1997).

p. 99, *federal campaign contributions*: Center for Responsive Politics (www.opensecrets.org).

p. 99, *they're doing business now*: Interview with Jim Saxton, May 18, 1999.

p. 99, *energy companies and alliances*: Lobbyist disclosure files and various corporate websites.

p. 100, *in deep-water leasing*: Interview with Robert Stewart, December 20, 1999. In an earlier interview, the author mentioned how writing about the ocean gets one to the beach. Stewart replied, "I don't like the beach."

p. 100, *some assistance to us*: Interviews with Hugh Depland, December 2–4, 1999.

p. 100, *New Orleans shopping mall*: Interviews with Jim Regg, December 9, 1999.

p. 102, *after Exxon-Mobil*: "Oil Merger Hung Up on Fields in Alaska," Associated Press story, *Times-Picayune*, December 3, 1999.

Chapter 6 A Rising Tide

p. 103, *northern winter of 1999*: The author traveled to Antarctica as part of a National Science Foundation program that allows several professional journalists to visit there each year.

p. 107, *60 million years ago*: On March 9, 2000, the Associated Press reported on a new scientific analysis, which found that after a species is extinct it takes 10 million years before anything resembling it reappears. The study "confirms the fears of many scientists, who estimate that half the Earth's species will be wiped out within a century."

p. 110, *impacts on Pacific islanders*: The PBS documentary *Rising Waters: Global Warming and the Fate of the Pacific Islands* was produced by Andrea Torrice. The tour took place on March 16, 1999.

p. 112, *of little storm activity*: Interview with Dr. Vivien Gornitz, May 3, 1999.

p. 113, *ghoulish ratings grabber*: The author's visit to the National Hurricane Center took place on July 26, 1999, following the widely reported search and recovery of John F. Kennedy, Jr.'s plane and the three bodies it contained.

p. 113, *the storm turned again*: Interviews with Deb Shephard, November 11 and 13, 1999.

p. 114, *from supersaturated graveyards*: Estimates of the number of hogs drowned ran from the North Carolina State Agriculture Department's low figure of 28,000 to the U.S. Department of Agriculture's estimate of more than 500,000.

p. 114, *like a big mix-master*: The highest storm tide mark, 16.9 feet above sea level, was also found on the side of the Burger King World Headquarters, according to NOAA's *Natural Disaster Survey Report, Hurricane Andrew: South Florida and Louisiana, August 23–26, 1999.*

p. 114, *more luck than anything*: Interview with Dr. Chris Landsea, July 14, 1999.

p. 115, *$100 billion category*: Steven Leatherman, appearing on the *CBS Evening News*, November 1, 1999.

p. 115, *refused to handle*: According to Mark Stevens, Office of Public Affairs, Federal Emergency Management Agency, April 6, 2000.

p. 115, *bearing down on the state*: Tanya Ott, "Marketplace" radio report, August 27, 1999. The lobbying continued through the following spring, with the home-builders insisting that only 4 of Florida's 36 counties needed hurricane-resistant building standards.

p. 116, *costing us lives*: U.S. Coast Guard, *Fiscal Year 2000 Budget in Brief*, and Admiral James M. Loy's address to the Center for Naval Analyses, December 1, 1999.

p. 117, *these situations very seriously*: Interview with Kathy Hersh, January 4, 2000.

p. 118, *$80 billion in damages*: Roger A. Pielke, Jr., and Christopher W. Landsea, "Normalized Hurricane Damages in the United States: 1925–1995," *Weather and Forecasting*, September 1998.

p. 119, *its master plan schematics*: According to interviews with Mike Davis, deputy assistant secretary of the army (civil works), and Stu Appelbaum, who oversees Corps of Engineers restoration work out of the Jacksonville office.

p. 119, *not counting Florida Bay*: Louisiana Coastal Wetlands Conservation and Restoration Task Force and the Wetlands Conservation and Restoration Authority, "Coast 2050," December, 1998.

p. 120, *five days that month*: Interview with Johnny Glover, December 8, 1999.

p. 120, *natural sediment transport*: "Coast 2050."

p. 120, *the next several decades*: A good overview of coastal problems and the "2050" response was written by *Baton Rouge Advocate* reporter Mike Dunne in the summer of 1999 and reprinted as a special report.

p. 120, *make the case for us*: Interviews with Mark Davis, November 17 and December 7, 1999.

p. 121, *commercially fished species*: Union of Concerned Scientists and the Ecological Society of America, *Confronting Climate Change in California* (1999), 46.

p. 121, *abandoned years ago*: Todd Shields, "Maryland Confronts Receding Shoreline," *Washington Post*, December 22, 1999.

Chapter 7 *Paradise with an Ocean View*

p. 124, *the rest of the country*: NOAA National Ocean Service Special Projects Office, "Trends in U.S. Coastal Regions, 1970–1998" (August 1999), 3.

p. 124, *It's called "newjerseyization"*: Orrin H. Pilkey and Katharine L. Dixon, *The Corps and the Shore* (Washington, D.C., Island Press, 1996), 42.

p. 125, *oversees the Corps of Engineers*: Interview with Dr. Joseph W. Westphal, January 21, 2000.

p. 125, *an erosion "hot spot"*: Michael Grunwald, "Whose Beaches, Whose Burdens?" *Washington Post,* April 20, 1999.

p. 126, *every seven years*: Interview with Dery Bennett (and House passage of WRDA), April 29, 1999.

p. 127, *a few big home owners*: Interview with Representative Jim Saxton, May 18, 1999.

p. 127, *beaches to their west*: Quoted in Pilkey and Dixon, *The Corps and the Shore,* 230–231.

p. 128, *the nation on coastal policy*: Thomas Maier, part four of the five-part series "Shoreline in Peril," *Newsday,* August 19, 1998.

p. 128, *continues on her way*: Interview with Jim O'Connell, May 5, 1999.

p. 129, *there's American sand available*: Interview with David Schmidt, July 19, 1999.

p. 129, *Environmental Protection Agency (EPA)*: Beth A. Millemann and Cindy Zipf, *Muddy Waters: The Toxic Wasteland Below America's Oceans, Rivers, and Lakes* (Washington, D.C.: Coast Alliance, Clean Ocean Action, and American Littoral Society, 1999); and NOAA reports.

p. 129, *the size of Manhattan*: Millemann and Zipf, *Muddy Waters,* 17–18; and newsclips and interviews.

p. 129, *chamber pot out the window*: Beth Millemann at a National Press Club Press conference, November 9, 1999.

p. 130, *net-loaded break-bulk cargo*: Stewart Taggart, "The 20-Ton Packet," *Wired,* October 1999; and Robert Mottley, "The Early Years," *American Shipper,* May 1996.

p. 131, *we have a soft bottom*: Interview with Frank Hamons, July 12, 1999. Even though his port had already lost the leases, he was having a hard time letting go.

p. 131, *projects set through 2004*: Lillian Borrone, "Time for Port Alliances?" *Container Management,* August 1999.

p. 132, *which are also expanding*: Gail Krueger, "Expert Says Ports Are 'Hostages,'" *Savannah Morning News,* November 10, 1999; interview with reporter Krueger; and her ongoing series for the *Morning News.*

p. 132, *the occasional philosophical possum*: This refers to the comic strip "Pogo," written and drawn by the late Walt Kelly. His Pogo possum character is often quoted from an Earth Day 1971 strip, in which, faced with a heavily polluted Pogofenokee Swamp, he says, "Yep, son. We have met the enemy and he is us!"

p. 133, *including personal watercraft*: NOAA National Ocean Service Special Projects Office, "Trends in U.S. Coastal Regions, 1970–1998" (August, 1999), 18–19.

p. 133, *in the United States*: It was called Cabbage Island at the time of the Williams sale. As it was developed for tourism, it was given the more visitor-friendly name Little Tybee Island.

p. 133, *we've got sand gnats*: Interview with Robert DeWitt, November 11, 1999.

p. 133, *the color of antifreeze*: Robert Sullivan, *The Meadowlands* (New York: Scribners, 1998), 18.

p. 134, *a doubling of acreage*: Language from Section 404B, Corps of Engineers guidelines established under the Clean Water Act of 1972; and discussions with corps officials and others.

p. 134, *has grown sympathetic*: Copies of correspondence between Baykeeper and other "environmental commentators" and Army Corps of Engineers, Environmental Protection Agency, White House Council on Environmental Quality, and Fish and Wildlife Service; and materials from newspapers and websites.

p. 134, *New Jersey Nets and Devils*: "A Meadowlands Quagmire: Development Plans Would Bring Headaches," *Bergen Record*, March 21, 1999.

p. 136, *a new city on its shore*: The author did extensive research for an article on Playa Vista that appeared in the March 1997 issue of *George* magazine. Among more recent articles, see Mark Hertsgaard, "Spielberg's Other Lost World," *Mother Jones*, January/February 1999.

p. 136, *How the hell does he know*: "Corps Gets Failing Report Card," *PEER Review*, summer 1999; "Peer, Corps Argue over Nationwide Wetlands Program," *Endangered Species and Wetlands Report*, August 1999; Scott Allen, "U.S. Pulls Back on Saving Marshes," *Boston Globe*, August 8, 1999; and interview with Jeff Rouch, January 6, 2000.

p. 136, *out there and investigating*: Interview with Dr. Joseph Westphal, January 21, 2000.

p. 137, *and Bob Smith*: Michael Grunwald, "Generals Push Huge Growth for Engineers," *Washington Post*, February 24, 2000; "Army Engineers Reforms Are Set," *Washington Post*, March 30, 2000; "Corps of Engineers Reforms Suspended," *Washington Post*, April 7, 2000, and other articles by Grunwald.

p. 137, *in flood insurance policies*: FEMA had $521,520,008,400 in flood insurance coverage as of April 6, 2000, according to Mark Stevens, FEMA Public Affairs Office.

p. 138, *fully mortgaged condominiums*: Cornelia Dean, *Against the Tide* (New York: Columbia University Press, 1999), 190.

p. 138, *factored into their rates*: Interview with Steve Ellis, January 3, 2000.

p. 138, *his trademark smirk for the camera*: 20/20 segment on federal flood insurance, produced by David Sloan, aired November 26, 1993.

p. 138, *Bogue Banks, Westhampton, and Malibu*: Nicholas Sparks, "I Will Rebuild," *New York Times*, September 19, 1999.

p. 138, *repetitive loss properties*: David Conrad, project director, *Higher Ground. National Wildlife Federation Report* (July 1998); and interview with Conrad.

p. 139, *between 1981 and 1996*: John Riley, part 3 of "Shoreline in Peril" series, *Newsday*, August 18, 1998, for Fire Island figures; Craig Whitlock, part 1 of "Flooded with Generosity," *Raleigh News and Observer*, November 9, 1997, for Topsail figures. Both took their data from FEMA.

p. 139, *Woods Hole's Jim O'Connell*: Interview with Jim O'Connell, May 5, 1999.

p. 139, *Peter Goss of Florida*: According to congressional staffers and copies of the testimony of three representatives before the Subcommittee on Fisheries, Conservation, Wildlife, and Oceans of the Committee on Resources.

p. 140, *appropriate for you to visit*: Telephone interview with Bob Berry, July 8, 1999.

p. 140, *by the [Clinton] administration*: Interview with Representative Peter Deutsch, June 30, 1999.

p. 140, *on Pumpkin Key*: Interview with Bob Berry and Tom Hayward, January 5, 2000.

p. 141, *one of our members*: Telephone interview with Bill Hackelton, July 23, 1999.

p. 141, *to assess the damage*: "J.S.R. Seaside Stands Firm as Hurricane Opal Wipes Out Stretches of Florida's Panhandle," *Architectural Record*, November 1995.

p. 142, *and by the water*: Interview with Robert Davis, January 10, 2000.

p. 143, *still ahead of us*: Interview with Representative Jim Saxton, May 18, 1999.

Chapter 8 Flushing the Coast

p. 147, *on a five-minute dive*: The author dove on Aquarius on July 16 and 17, 1999.

p. 148, *sugar growers in Florida*: From *The Starr Report: The Findings of Independent Counsel Kenneth W. Starr on President Clinton and the Lewinsky Affair. With Analysis by the staff of the Washington Post* (New York: Public Affairs, 1998).

p. 148, *in that election*: Various clippings, including Dwight L. Morris, "Playing It Safe," Washington Post.com, June 10, 1996.

p. 149, *between 1960 and 1990*: Peter M. Vitousek et al., "Human Alteration of the Global Nitrogen Cycle: Sources and Consequences," Technical Report in *Issues in Ecology* 7, no. 3 (1997): 737–750; Vaclav Smil, "Global Population and the Nitrogen Cycle," *Scientific American*, July 1997. The 1960–1990 figure is from National Research Council, *Clean Coastal Waters: Understanding and Reducing the Effects of Nutrient Pollution* (Ocean Studies Board, 2000).

p. 149, *on an ocean planet*: Interviews with Steve Miller, July 15–18, 1999.

p. 151, *another close one*: Interview with Craig Cooper, July 16, 1999.

p. 152, *fertilizer per acre*: Bill Lambrecht, "Fishers Want Farmers to Be More Responsible," *St. Louis Post-Dispatch*, August 24, 1997.

p. 153, *a conservative think-tank author*: Michael Fumento, "Hypoxia Hysteria," *Forbes*, November 15, 1999.

p. 153, *heavily invested in agrochemicals*: For more on Farm Bureau, see Helvarg, *The War Against the Greens* (paper ed., 1997), 25–28, 52–53, 320–321.

p. 154, *but it's going to happen*: Interview with Jonathan Pennock, December 6, 1999.

p. 155, *pollution of U.S. waters*: National Research Council, *Clean Coastal Waters: Understanding and Reducing the Effects of Nutrient Pollution* (Ocean Studies Board, 2000); and the accompanying National Academies press release, April 4, 2000.

p. 155, *didn't have any answers for him*: Interview with Nancy Rabalais, December 8, 1999.

p. 156, *in October 1999*: Peter Annin, "Down in the Dead Zone," *Newsweek*, October 18, 1999.

p. 156, *debilitating neurological disorders*: From Rodney Barker, *And the Waters Turned to Blood* (New York: Simon & Schuster, 1997).

p. 156, *rains out into the water*: Phil Bowie, "No Act of God," *Amicus Journal*, winter 2000.

p. 157, *a 1996 Pulitzer Prize*: The series "Boss Hog," written by Pat Stith and Joby Warrick, ran from February 26 to March 4, 1995, in *Raleigh News and Observer*.

p. 158, *second largest poultry company*: From Peter S. Goodman, "Poultry's Price/ The Cost to the Bay," *Washington Post*, August 1–3, 1999.

p. 158, *legally requires them to use*: Ibid., part 3, "Who Pays for What Is Thrown Away?"

p. 158, *DO levels of 1 part per liter*: From multiple sources, including Associated Press, "Dead Zone Threatens Carolina," *Marin Independent Journal*, October 9, 1999; and series by James Shiffer including "Forces of Nature and Man," *Raleigh News and Observer*, November 7, 1999.

p. 158, *on the Blue Frontier*: Woods Hole Oceanographic Institution, *Ecohab: The Ecology and Oceanography of Harmful Algal Blooms. A National Research Agenda* (Woods Hole, Mass., Woods Hole Oceanographic Institution, 1995).

p. 159, *irritations for beachgoers*: From *Dallas Morning News*, October 5, 1997; *Washington Post*, September 23, 1997; and *New York Times*, January 10, 2000. Additional stories are listed on the Woods Hole Harmful Algal Blooms website (habserv1.whoi.edu/hab).

p. 159, *the stinging jellies*: From interviews at Dauphin Island Sea Lab, December 5–6, 1999, and CBS *Evening News*, "A Stinging Sign?" January 12, 2000.

p. 159, *buy my oysters*: The agent Rich Severtson, a kind of "seafood Serpico," ran undercover operations for NOAA law enforcement out of the Seattle National Marine Fisheries Service offices.

p. 160, *their beach businesses hammered*: "Perspectives," *Newsweek*, September 13, 1999.

p. 160, *for children and adults*: Charles J. Carter, "'Surf City' Is Riding Wave of Despair," Associated Press, reprinted in *SD Union-Tribune*, August 28, 1999; David Reyes and Louise Roug, "Beach Reopened after Needle Cleanup," *Los Angeles Times*, September 18, 1999; and David Helvarg, "Congress Plans an American Clearcut," *The Nation*, December 4, 1995.

p. 160, *those sections of the beach*: Robert W. Haile et al., "An Epidemiological Study of Possible Adverse Effects of Swimming in Santa Monica Bay," Santa Monica Bay Restoration Project, May 7, 1996.

p. 161, *That's just a joke*: From Judy Wilson's panel talk at the Society of Environmental Journalists Conference, University of California at Los Angeles, September 17, 1999; and interview with Wilson, January 28, 2000.

p. 163, *capable of regeneration for itself*: Interview with Senator John Kerry, January 21, 2000.

Chapter 9 The Last Fish

p. 165, *Department of Environmental Conservation (DEC)*: Fulton Fish Market visit with National Marine Fisheries Service, May 3, 1999. A friend, New York high school teacher Mark Ambrosino, came along.

p. 166, *illegally caught fish*: Eric Lipton, "5 Dealers Charged in Sale of Bass from Polluted Waters," *New York Times*, December 9, 1999, and other clips.

p. 167, *from Ingold and other fishermen*: Al Guart, "5 Hooked in Toxic-Fish Sales Ring," *New York Post*, December 9, 1999; and Greg Smith, "Bad Fish on Fancy Menus," *New York Daily News*, December 9, 1999.

p. 167, *dealing in unreported fish*: From a copy of the two-page NOAA general counsel's administrative fine notice dated December 9, 1998, and signed by J. Mitch MacDonald, NOAA enforcement attorney; and follow-up conversation with NMFS agent James McDonald.

p. 168, *the wall of the dockhouse*: Copies of National Marine Fisheries Service offense investigation reports filed with NOAA legal staff on September 24, 1997, and May 19, 1998.

p. 170, *booming aquaculture industry*: *PEER Review*, Spring 2000, 10.

p. 170, *and other trawlers fined*: U.S. Department of Commerce press release dated May 27, 1999, *National Fisherman*, August 1999, 8–9; and follow-up calls to NMFS.

p. 170, *status is unknown to them*: Interview with Andy Rosenberg, June 1, 1999. NMFS, Report to Congress: Status of Fisheries off the United States, October 1999.

p. 171, *faster than that of land animals*: *Conservation Biology*, October 1999; also referenced by Representative Sam Farr in *Sea Technology* magazine, January 2000, 19.

p. 172, *my life by his choice*: The pursuit and interviews took place in early 1997 as the author was preparing an article on National Marine Fisheries Service enforcement agents for *Smithsonian* magazine. Pete Choerny has since retired.

p. 172, *funding for TEDs research*: The author went out on the *Bulldog*, a research shrimper owned by the University of Georgia, on November 11, 1999. For one political history of TEDs, see Center for Marine Conservation, *Delay and Denial* (Washington, D.C.: Center for Marine Conservation. 1995).

p. 173, *in order to protect fish*: Interviews with Zeke Grader, April 14 and September 3, 1999, and January 9, 2000.

p. 173, *Louisiana, Idaho, and Appalachia*: National Oceanographic and Atmospheric Administration, *Trends in U.S. Coastal Regions, 1970–1998* (Washington, D.C.: National Ocean Service Special Projects Office, August 1999), 11–12.

p. 174, *escape into the wild*: Visit to salmon farm, East Johnson Bay, Maine, July 23, 1998; Rebecca Goldburg and Tracy Triplett, *Murky Waters: Environmental Effects of Aquaculture in the United States* (Washington, D.C.: EDF Publications, 1997); and reports from Sea Grant Program and National Fisherman.

p. 174, *a cobblestone street*: C. J. Chivers, "Scraping Bottom," *Wildlife Conservation*, February 2000.

p. 174, *into simple cow pastures*: Sue Robinson, "The Battle Over Bottom Trawling," *National Fisherman*, August 1999; and "Scraping Bottom," Marine Conservation Biology Institute press packet and news release, December 14, 1998.

p. 175, *on returning to port*: From the foreword of William W. Warner, *Distant Water: The Fate of the North Atlantic Fisherman* (New York: Penguin Books, 1997).

p. 176, *how to do it better*: Strong-Cevetich is the founder of SEACOPS (Southeast Alaska Coalition Opposed to the Piracy of Salmon), which fought for the banning of high-seas drift nets in the early 1990s. Quote is from 1992.

p. 176, *Eat Fish Twice a Week*: *Federal Fisheries Investment Task Force Report to Congress*, July 1999, Executive Summary, p. 27.

p. 177, *to get into the industry*: Interview with Paul Cohan, May 26, 1999.

p. 178, *they can do it, too*: Interview with Rod Avila, May 4, 1999.

p. 178, *pick up the tab in the end*: Brad Matsen, "Picking Up the Groundfish Tab," *National Fisherman*, October 1992.

p. 178, *which was fishing*: Interview with Billy Causey, July 19, 1999.

p. 179, *pulls into port*: The author interviewed Dan O'Brian several times while preparing stories on fisheries enforcement. This material is from 1995.

p. 180, *three times a year*: From New England Fishery Management Council transcript for May 26, 1999, and author's notes. The councilman was Doug Hopkins.

p. 182, *Yozell responds meekly*: Senate hearing on marine sanctuaries and coral reefs, June 30, 1999.

p. 182, *what's best for the resource*: Interview with Rod Avila, May 4, 1999.

p. 182, *little incentive for a buyout*: Interview with Jim Saxton, May 18, 1999.

p. 182, *divided the loot evenly*: National Oceanographic and Atmospheric Administration, *Magnuson-Stevens Fishery Conservation and Management Act, Title 3, Voting Members, and Disclosure of Financial Interest and Recusal.* (Washington, D.C.: NOAA Technical Memo, December 1996).

p. 183, *to scallop dredging*: Copy of letter from William M. Daley to Joseph Brancaleone, Chairman, New England Fishery Management Council, February 23, 1999.

p. 183, *explains the fund's Bob Bruno*: Interview with Bob Bruno, May 4, 1999.

p. 183, *every pound of scallops*: Scott Allen, "Return to Georges Bank," *Boston Globe,* April 5, 1999; and video "Back Against the Wall," produced for the Fisheries Survival Fund.

p. 183, *no longer being available*: Copies of correspondence between Massachusetts politicians (including Kerry) and Secretary Daley and interviews with various Kerry staffers, fishing organizations, and National Marine Fisheries Service.

p. 184, *other kind of work myself*: Interview with Andy Philips, September 3, 1996.

p. 184, *things might change*: Interview with Sylvia Earle, July 18, 1999.

p. 185, *missing the big picture*: Interview with Zeke Grader, February 8, 2000.

p. 186, *Pacific Fisheries Council*: Dr. Walter Pereyra's testimony at Panel 1, Oversight Hearing on Steller Sea Lions. House Committee on Resources, May 20, 1999. Pereyra has since retired from the council vice-chairmanship.

Chapter 10 Drowning in Red Tape

p. 187, *Johnson and Nixon*: Interviews with Edward Wenk in 1995 and on November 22, 1999.

p. 187, *of the marine environment*: Edward Wenk, Jr., *The Politics of the Ocean* (Seattle ,Wash.: University of Washington Press. 1972), 359.

p. 188, *protest must be heard*: Walter J. Hickel, *Who Owns America?* (Englewood Cliffs, N.J.: Prentice-Hall, 1971), 247–249.

p. 189, *win her that supremacy*: Wenk, *Who Owns America?*, 52.

p. 189, *must help us meet*: Preface by Warren G. Magnuson to E. John Long, *New Worlds of Oceanography* (New York: Pyramid, 1965), 18.

p. 189, *respirating through artificial gills*: Senator Claiborne Pell with Leland

Goodwin, *Challenge of the Seven Seas,* (New York: Morrow, 1966), 1–24.

p. 190, *a wet NASA:* The Stratton Roundtable, National Ocean Service, NOAA, and Delaware Sea Grant College Program, 1998 and various books and interviews.

p. 191, *the Weather Service:* Wenk, *The Politics of the Ocean,* 360.

p. 191, *with the oceans:* Interview with Robert White, March 2, 2000.

p. 192, *icon of Northwest wilderness:* Jim Lichatowich, *Salmon Without Rivers* (Washington, D.C.: Island Press, 1999); David James Duncan, "Salmon's Second Coming," *Sierra* magazine, March/April 2000; and various news reports.

p. 194, *seems to know this:* Copy of Presidential Decision Directive/NSC-36, White House, April 5, 1995.

p. 194, *wake-up call from nature:* Interview with Sylvia Earle, July 18, 1999.

p. 194, *nothing happens:* Interview with Roger McManus, March 31, 1999.

p. 195, *on resource protection:* Interview with Zeke Grader, February 8, 2000.

p. 195, *the factory trawler industry:* Joel Gay, "The Will to Win." *National Fisherman,* December 1999; and David Helvarg, "Full Nets, Empty Seas," *The Progressive,* November 1997.

p. 196, *Anchorage attorney William Bittner:* David Whitney, "Stevens' Factory Ship Bill Sets Off Lobbying Frenzy," *Anchorage Daily News,* March 17, 1998 (from newpaper's website, adn.com).

p. 196, *a considerable raise in pay:* From interviews with former Stevens staffers, NOAA officials, Greenpeace and other lobbyists, and At-Sea Processors Association (www.atsea.org).

p. 196, *Russian side of the pollack line:* PA2 Edwin Lyngar, PacArea, "Dicey Icy Rescue," *Coast Guard* magazine, March 1999.

p. 197, *of nonrenewable resources:* Coastal Impact Assistance Working Group, *Report to the OCS Policy Committee,* undated copy; and Mineral Management Service, *Moving Beyond Conflict to Consensus. The Outer Continental Shelf Oil and Gas Program* (October 1993).

p. 199, *time for me to leave, too:* Senate Committee on Energy and Natural Resources hearing on April 20, 1999.

p. 199, *with a bulldozer:* Surfrider Clean Water Paddle Out, Ocean Beach Pier, August 28, 1999.

p. 200, *his own political agenda:* Interview with Donna Frye, August 28, 1999.

p. 200, *still is in San Diego:* David Helvarg, "Congress Plans an American Clearcut," *The Nation,* December 4, 1995.

p. 201, *they would not pass:* *Congressional Record,* H4760. May 10, 1995: "The question was taken, and the Chairman announced that the no's appeared to have it. Mr. Saxton, Mr. Chairman. I demand a recorded vote. A recorded vote was ordered. Mr. Bilbray changed his vote from 'no' to 'aye.'"

p. 201, *input we want on it is fine*: Interview with Barbara Jeanne Polo, April 13, 1999.

p. 202, *pumps into the ocean*: Steve La Rue and Terry Rodger, *San Diego Union* articles February 29–March 3, 2000, starting with "Massive Sewage Spill Was Long Undetected" and ending with "City Files Lawsuit over EPA Warning," from *Union* website (signonsandiego.com).

p. 202, *big polluters in charge*: Interview with Tom Sekreta, August 28, 1999.

p. 203, *New Zealand, and elsewhere*: Interview with Will Travis, September 10, 1999, and Bay Conservation and Development Commission materials.

p. 204, *did a 180-degree turn*: Interview with Peter Douglas, September 1, 1999.

p. 204, *we needed Prop 20*: Interview with Representative Sam Farr, February 4, 2000.

p. 205, *Wheeler and other principals*: "The Coast's Best Friend," *Sacramento Bee*, July 12, 1996; "The Political Game Is On, and Coastline Is the Loser," *Los Angeles Times*, July 8, 1996; and numerous other news and editorial clips.

p. 206, *vocal special interest groups*: Jeffrey I. Rabin and Deborah Schochi, "Coastal Commission Halts Bid to Fire Director," *Los Angeles Times*, July 13, 1996; and Alex Barnum, "Coastal Chief Keeps His Job—For Now," *San Francisco Chronicle*, July 13, 1996.

p. 206, *run for the governorship*: Interview with Sara Wan, September 9, 1999; other interviews; and news clips.

p. 207, *ocean fish and marine mammals*: Election returns and proposition language from *San Francisco Chronicle* website (sfgate.com), March 9, 2000.

Chapter 11 Sanctuaries in the Sea

p. 210, *restocking these animals*: The author went out on patrol with officer Greg Stanley on July 28, 1999.

p. 210, *residents now support increased protection*: United States Coral Reef Task Force, *Coastal Uses Working Group Summary Report*, March 2, 2000.

p. 212, *towing a Jet Ski*: Statistics from Department of Transportation brochure, "Our Valuable U.S. Marine Transportation System."

p. 212, *I take a ride with*: Patrol and interview with Dave McDaniel, July 21, 1999.

p. 213, *is an open question*: The Marine Protection, Research, and Sanctuaries Act of 1972 established a regulatory framework for ocean dumping in U.S. waters along with the sanctuary program itself.

p. 214, *Newport News, Virginia*: Bruce G. Terrell, *Fathoming Our Past* (Newport News, Va.: Mariners' Museum, NOAA), 31–32; Larabee et al., *America and the Sea*, 353; *Wild Oceans*, 194–195; and various interviews.

p. 214, *protection of the coastline*: The author did extensive research on the sanctuaries' history for the cover article "Blue Frontiers," *Audubon,* June 1995.

p. 215, *the mouse that roared*: Interview with Sam Farr, February 4, 2000.

p. 215, *It was real groundswell*: Interview with Leon Penetta in 1995 for "Blue Frontiers," *Audubon,* June 1995.

p. 216, *she wondered*: Author's notes from Sunday, September 20, 1992.

p. 218, *their population decline*: Todd Wilkinson, "Marine Mystery," *National Parks* magazine, March/April 2000; Friends of the Sea Otter website (seaotters.org); and interviews with biologists.

p. 218, *endangered coral reef life*: For a first-person description of spawning coral, see Douglas H. Chadwick, "Blue Refuges," *National Geographic,* March 1998.

p. 219, *animals are highly intelligent*: Observations from the *Gustavus* ferry based out of Auke Bay, north of Juneau, Alaska, August 27, 1998.

p. 219, *set aside as parks and wilderness*: Helvarg, *The War Against the Greens,* 442.

p. 219, *have their own boats*: Interview with Chris Ostrum, March 10, 2000. Dollar figures are from NOAA Sanctuaries Office.

p. 220, *any other national park*: NOAA National Ocean Service Special Projects Office, *Trends in U.S. Coastal Regions, 1970–1998* (August 1999), 7.

p. 220, *the oceangoing public*: Interview with National Park Service marine scientist Gary Davis, March 21, 2000.

p. 220, *if we let this happen*: Interview with Sylvia Earle, July 18, 1999.

p. 220, *within the sanctuaries*: Interview with Steve Giddings, science coordinator for the National Marine Sanctuaries program, March 17, 2000; and Sustainable Seas Expedition materials.

p. 221, *Humpback Whale Sanctuary in Hawaii*: Project Jason press kit and website (www.jasonproject.org).

p. 222, *stalking their prey from below*: Interview with Ken Goldman, 1996, and other shark information from notes for author's article, "Great White Comeback," *Men's Journal,* June/July 1996. Goldman is now at the Virginia Institute of Marine Science.

p. 224, *natural cycles of the sea*: Interviews with Ed Ueber in 1991, 1995, 1997, and on September 13, 1999.

p. 224, *what the public wants*: Interview with Brad Barr, May 6, 1999.

p. 224, *program for environmental reporters*: Institute for Journalists and Natural Resources is run out of Montana by former *Wall Street Journal* reporter and editor Frank Allen and his wife, Maggie. Its aim is to improve reporting of natural resource issues and conflicts.

p. 225, *and daily ferry*: The Sapelo Island visit took place on November 12, 1999.

Chapter 12 The Seaweed Rebellion

p. 229, *$10,000-a-year refuge budget*: In February 2000 the Refuge office moved to a double-wide trailer on Route 905.

p. 230, *in my own little world*: Interviews with Steve Klett, July 21–22, 1999, and March 24, 2000.

p. 233, *their once scenic shoreline*: Geoff Pender, "Fury Along the Sound, SAND Tries to Unite Activists," *Biloxi Sun Herald*, December 5, 1999.

p. 234, *and swordfish*: William J. Broad, "Conservationists Write a Seafood Menu to Save Fish," *New York Times*, November 9, 1999.

p. 234, *and Alaskan wild salmon*: Jay Lindsay, Associated Press, "Seafood Seal, New Label Indicates Environmentally Friendly Seafood," March 9, 2000; and Fish Forever press kit.

p. 234, *fisheries to sustain themselves*: Interview with Roger Berkowitz, March 31, 2000.

p. 236, *TV stations this evening*: The sharks' transport and release took place July 21 and 22, 1999, and was widely covered by Chicago and south Florida media, including Miami channels 4 and 7.

p. 237, *bites him in the foot*: Avery Sumner, "Tourist Bit by Bull Shark," *Key West Citizen*, July 28, 1999.

p. 237, *in San Jose, California*: Interview with Rick Trout, February 11, 2000.

p. 237, *South Florida Water District*: Dinner was on July 26, 1999. Other Reef Awareness Week events included a film festival, rope-splicing party, science forums, and sunset cruises.

p. 238, *dead zones along the coast*: Interviews with Ray Vaughan in 1997 and 1998, on December 5 and 6, 1999; and on March 15, 2000.

p. 239, *26-foot research vessels*: Interview with George Crozier, December 6, 1999. The boat was operated by Dr. John Dindo, also of Dauphin Island Sea Lab.

p. 239, *above its assessed value*: Crozier, who sits on the board of Forever Wild, the state agency that negotiated the deal, was not allowed to give me the assessed value of the land but said it was "much, much less" than the asked-for $20 million.

p. 239, *conservation and natural resources*: Copy of 1996 legal complaint, "Forever Dauphin Island vs. Alabama Department of Environmental Management and Alabama Environmental Management Commission." Alabama's Democratic Governor Don Siegelman named Smith the director of Conservation and Natural Resources.

p. 240, *a Hawaiian hula group*: San Francisco Surfrider Clean Water Paddle Out and Luau, August 21, 1999.

p. 241, *They've got to watch*: Interview with Doug Hartley, August 21, 1999.

p. 241, *identifying ships, I think*: Interview with Dery Bennett, April 29, 1999.

p. 242, *as large as bulldogs*: Charles Gaines, "The Love Song of the Linesiders," *Sports Afield*, April 2000.

p. 244, *incentive for illegal poaching*: Elver fishermen discard catch in protest of price—eels fetch $20 per pound, down from $300. From *National Fisherman*, August 1999.

p. 245, *security and pollution*: Interview with Representative Curt Weldon, March 30, 2000.

p. 245, *windsurf, bodyboard, or surf kayak*: NOAA National Ocean Service Special Projects Office, *Trends in U.S. Coastal Regions, 1970–1998* (August 1999), 7; and various other reports.

p. 245, *100 percent of the moon*: From a speech by Admiral Paul Gaffney, March 8, 2000.

p. 245, *my angel, my kitten*: Michael Cabanatuan and Carolyne Zinko, "Couple's Honeymoon Tragedy in Huge Tempest-Tossed Seas," *San Francisco Chronicle*, October 30, 1999.

p. 247, *other self-conscious stylists*: Robert Hughes, *American Visions* (New York: Knopf, 1999), 303–316; Larabee et al., *American and the Sea*, 396–397; and various exhibits and prints.

Bibliography

Alic, John A., Lewis M. Branscomb, Harvey Brooks, Ashton B. Carter, and Gerald L. Epstein. 1992. *Beyond Spinoff: Military and Commercial Technologies in a Changing World*. Boston: Harvard Business School Press.

Ballard, Robert D., with Will Hively. 2000. *The Eternal Darkness*. Princeton, N.J.: Princeton University Press.

Barker, Rodney. 1997. *And the Waters Turned to Blood*. New York: Simon & Schuster.

Barlow, Jeffrey G. 1994. *Revolt of the Admirals*. Washington, D.C.: Dept. of the Navy.

Bascom, Willard. 1988. *The Crest of the Wave*. New York: Harper & Row.

Benchley, Peter, and Judith Gradwohl. 1995. *Ocean Planet*. New York: Abrams.

Berendt, John. 1994. *Midnight in the Garden of Good and Evil*. New York: Random House.

Berrill, Michael. 1997. *The Plundered Seas*. San Francisco, Calif.: Sierra Club Books.

Biel, Steven. 1996. *Down With the Old Canoe*. New York: Norton.

Borgese, Elisabeth Mann, ed. 1992. *Ocean Frontiers*. New York: Abrams.

Boyer, Richard O., and Herbert M. Morals. 1997. *Labor's Untold Story*. Pittsburgh, Pa.: Pittsburgh United Electrical Radio and Machine Workers of America.

Broad, William J. 1997. *The Universe Below*. New York: Simon & Schuster.

Brooke, Steven. 1995. *Seaside*. Gretna, La.: Pelican Publishing.

Brower, Kenneth. 1991. *Realms of the Sea*. Washington, D.C.: National Geographic.

Burleson, Clyde W. 1997. *The Jennifer Project*. College Station, Tex.: Texas A&M University Press.

California Coastal Commission. 1987. *California Coastal Resource Guide*. Berkeley, Calif.: University of California Press.

Carey, Richard Adams. 1999. *Against the Tide: The Fate of the New England Fisherman*. Boston, Mass.: Houghton Mifflin.

Carson, Rachel L. 1951. *The Sea Around Us*. New York: Oxford University Press.

Cicin-Sain, Biliana, and Robert W. Knect. 2000. *The Future of U.S. Ocean Policy*. Washington, D.C.: Island Press.

Cicin-Sain, Biliana, Robert W. Knect, and Nancy Foster. 1999. *Trends and Future Challenges for U.S. National Ocean and Coastal Policy*. Washington, D.C.: Dept. of Commerce/NOAA.

Clarke, Arthur C. 1957. *The Deep Range*. New York: Harcourt, Brace.

Coastal Zone Management. 1974. "The Coastal Imperative: Developing a National Perspective for Coastal Decision Making." Senate Committee on Commerce, Washington, D.C.: Government Printing Office.

Committee on Interior and Insular Affairs, U.S. House of Representatives, 102nd Congress. 1992. *Alyeska Pipeline Service Company Covert Operation*. Washington, D.C.: Government Printing Office.

Conrad, David R., Ben McNitt, and Martha Stout. 1998. *Higher Ground*. Washington, D.C.: National Wildlife Federation.

Cousteau, Jacques Yves. 1953. *The Silent World*. New York: Harper.

———. 1963. *The Living Sea*. New York: Harper & Row.

Cousteau, Jean-Michel, and Mose Richards. 1992. *Cousteau's Great White Shark*. New York: Abrams.

Cox, Donald W. 1968. *Explorers of the Deep: Pioneers of Oceanography*. Maplewood, N.J.: Hammond.

Cronin, John, and Robert F. Kennedy, Jr. 1997. *The Riverkeepers*. New York: Simon & Schuster.

Cuyvers, Luc. 1993. *Sea Power*. Annapolis, Md.: Naval Institute Press.

Davis, Chuck. 1991. *California Reefs*. San Francisco: Chronicle Books.

Davis, Richard A., Jr. 1997. *The Evolving Coast*. New York: Scientific American Library.

Department of the Navy. 1999. *Vision, Presence, Power*. Washington, D.C.: Government Printing Office.

DeWitt, John. 1999. *Protecting Our National Marine Sanctuaries*. Washington, D.C.: National Academy of Public Administration.

Dorfman, Mark. 1999. *Testing the Waters*. New York: Natural Resources Defense Council.

Doubilet, David. 1989. *Light in the Sea*. Charlottesville, Va.: Thomasson-Grant.

Duane, Daniel. 1996. *Caught Inside: A Surfer's Year on the California Coast*. New York: Farrar, Straus.

Earle, Sylvia A. 1995. *Sea Change*. New York: Putnam.

Earle, Sylvia A., and Al Giddings. 1980. *Exploring the Deep Frontier*. Washington, D.C.: National Geographic.

Earle, Sylvia A., and Henry Wolcott. 1999. *Wild Oceans: America's Parks Under the Sea*. Washington, D.C.: National Geographic.

Eaton, John P., and Charles A. Haas. 1995. *Titanic: Triumph and Tragedy*. New York: Norton.

Ellis, Richard, and John E. McCosker. 1991. *Great White Shark*. Stanford, Calif.: Stanford University Press.

Environmental Health Center. 1999. *Covering Key Environmental Issues*. Washington, D.C.: Radio and Television News Directors Foundation.

Epstein, Richard A. 1985. *Takings: Private Property and the Power of Eminent Domain*. Cambridge, Mass.: Harvard University Press.

Federal Fisheries Investment Task Force. 1999. *Report to Congress*. Washington, D.C.: National Oceanographic and Atmospheric Administration.

Fisher, David E. 1994. *The Scariest Place on Earth: Eye to Eye with Hurricanes*. New York: Random House.

Fisheries Statistics and Economics Division. 1999. *Fisheries of the United States, 1998*. Washington, D.C.: Government Printing Office.

Fordham, Sonja V. 1996. *New England Groundfish: From Glory to Grief*. Washington, D.C.: Center for Marine Conservation.

Foundation for American Communications. 1995. *Reporting on Oceans*. Los Angeles, Calif.: Foundation for American Communications.

Friedheim, Robert L. 1993. *Negotiating the New Ocean Regime*. Columbia, S.C.: University of South Carolina Press.

Goldburg, Rebecca, and Tracy Triplett. 1997. *Murky Waters: Environmental Effects of Aquaculture in the United States*. Washington, D.C.: EDF Publications.

Gramling, Robert. 1996. *Oil on the Edge*. Albany, N.Y.: State University of New York Press.

Greenlaw, Linda. 1999. *The Hungry Ocean*. New York: Hyperion.

Greider, William. 1998. *Fortress America*. New York: Public Affairs.

Griffin, M. D., and L. Martin. 1998. *Saving the Marin-Sonoma Coast*. Healdsburg, Calif.: Sweetwater Springs Press.

H. John Heinz Center for Science, Economics, and the Environment. 1999. *Designing a Report on the State of the Nation's Ecosystems*. Washington, D.C.: Heinz Center.

———. 1999. *The Hidden Costs of Coastal Hazards*. Washington, D.C.: Island Press.

Hagan, Kenneth J. 1991. *This People's Navy.* New York: Free Press.

Halberstadt, Hans. 1999. *U.S. Navy SEALS.* New York: Barnes & Noble.

Hamilton-Paterson, James. 1992. *The Great Deep.* New York: Henry Holt.

Hansen, Gunnar. 1993. *Islands at the Edge of Time.* Washington, D.C.: Island Press.

Harrigan, Stephen. 1992. *Water and Light.* San Francisco, Calif.: Sierra Club Books.

Hendrickson, Robert. 1984. *The Ocean Almanac.* New York: Doubleday.

Hersey, John. 1994. *Key West Tales.* New York: Knopf.

Hiaasen, Carl. 1995. *Stormy Weather.* New York: Knopf.

———. 2000. *Sick Puppy.* New York: Knopf.

Hickel, Walter J. 1971. *Who Owns America?* Englewood Cliffs, N.J.: Prentice-Hall.

Hofstadter, Richard, and Michael Wallace, eds. 1971. *American Violence: A Documentary History.* New York: Vintage.

Holing, Dwight. 1990. *Coast Alert.* Washington, D.C.: Island Press.

Howarth, Stephen. 1991. *To Shining Sea.* Norman, Okla.: University of Oklahoma Press.

Hughes, Robert. 1999. *American Visions: The Epic History of Art in America.* New York: Knopf.

Idyll, C. P. 1970. *The Sea Against Hunger.* New York: Thomas Y. Crowell.

Iudicello, Suzanne, Michael Weber, and Robert Wieland. 1999. *Fish, Markets, and Fishermen.* Washington, D.C.: Island Press.

Jacobs, John. 1995. *A Rage for Justice: The Passion and Politics of Phillip Burton.* Berkeley, Calif.: University of California Press.

Jasny, Michael. 1999. *Sounding the Depths.* New York: Natural Resources Defense Council.

Johnson, Robert Erwin. 1987. *Guardians of the Sea: History of the United States Coast Guard.* Annapolis, Md.: Naval Institute Press.

Junger, Sebastian. 1997. *The Perfect Storm.* New York: Norton.

Kleinberg, Howard. 1989. *Miami: The Way We Were.* Surfside, Fla.: Surfside Publishing.

Koplow, Douglas, and Aaron Martin. 1998. *Fueling Global Warming.* Washington, D.C.: Greenpeace.

Kunzig, Robert. 1999. *The Restless Sea.* New York: Norton.

Kurlansky, Mark. 1997. *Cod.* New York: Penguin Books.

Labaree, Benjamin W., et al. 1998. *America and the Sea*. Mystic, Conn.: Mystic Seaport.

Lancek, Lena, and Gideon Bosker. 1998. *The Beach: The History of Paradise on Earth*. New York: Viking.

Larson, Erik. 1999. *Isaac's Storm*. New York: Crown.

Leary, William M. 1999. *Under Ice: Waldo Lyon and the Development of the Arctic Submarine*. College Station, Tex.: Texas A&M University Press.

Lewis, Charles, and the Center for Public Integrity. 2000. *The Buying of the President 2000*. New York: Avon Books.

Lichatowich, Jim. 1999. *Salmon without Rivers*. Washington, D.C.: Island Press.

Lilly, John C. 1961. *Man and Dolphin*. New York: Doubleday.

London, Jack. 1905. *Tales of the Fish Patrol*. London: Macmillan.

Long, E. John. 1965. *New Worlds of Oceanography*. New York: Pyramid.

Maas, Peter. 1999. *The Terrible Hours*. New York: Harper Collins.

Maier, Pauline. 1974. *From Resistance to Revolution*. New York: Vintage Books.

Marine Board National Research Council. 1996. *Undersea Vehicles and National Needs*. National Academy Press.

Marx, Wesley. 1999. *The Frail Ocean: A Blueprint for Change in the New Millennium*. Chester, Conn.: Globe Pequot Press.

Mathews-Amos, Amy, and A. Ewann Berntson. 1999. *Turning Up the Heat: How Global Warming Threatens Life in the Sea*. Washington, D.C.: World Wildlife Fund.

McComb, David G. 1986. *Galveston: A History*. Austin, Tex.: University of Texas Press.

McCullough, David. 1977. *The Path between the Seas*. New York: Simon & Schuster.

McGinn, Anne Platt. 1999. *Safeguarding the Health of Oceans*. Washington, D.C.: Worldwatch Institute.

McPhee, John. 1976. *Coming into the Country*. New York: Farrar, Straus.

———. 1990. *Looking for a Ship*. New York: Farrar, Straus.

MEDEA. 1995. *Scientific Utility of Naval Environmental Data*. McLean, Va.: MEDEA.

———. 1997. *Ocean Dumping of Chemical Munitions: Environmental Effects in Arctic Seas*. McLean, Va.: MEDEA.

Melville, Herman. 1851. *Moby-Dick, or The Whale*. New York: Harper & Brothers. New York: Penguin Classics, 1992.

Mileti, Dennis S. 1999. *Disasters by Design*. Washington, D.C.: Joseph Henry Press.

Millemann, Beth, and Cindy Zipf. 1999. *Muddy Waters: The Toxic Wasteland Below America's Oceans, Rivers, and Lakes.* Washington, D.C.: Coast Alliance, Clean Ocean Action, American Littoral Society.

Miller, Nathan. 1992. *Stealing From America.* New York: Paragon House.

Morgan, Judith and Neil. 1996. *Roger.* San Diego, Calif.: Scripps Institution of Oceanography.

Mowrey, Marc, and Tim Redmond. 1993. *Not in Our Backyard.* New York: William Morrow.

National Marine Fisheries Service. October 1999. *Report to Congress: Status of Fisheries of the United States.*

National Oceanographic and Atmospheric Administration. December 1996. NOAA Technical Memorandum NMFS-F/SPO-23: Magnuson-Stevens Fishery Conservation and Management Act. Washington, D.C.: Government Printing Office.

National Oceanographic and Atmospheric Administration. 1999. NOAA Technical Memorandum NMFS-F/SPO-41. *Our Living Oceans.* Washington, D.C.: Government Printing Office.

National Oceanographic Program. 1969. *Hearings before the Subcommittee on Oceanography of the Committee on Merchant Marine and Fisheries, House of Representatives.* Parts 1 and 2. Washington, D.C.: Government Printing Office.

National Research Council. 1997. *Oceanography and Naval Special Warfare.* Washington, D.C.: National Academy Press.

National Weather Service. November 1993. *Hurricane Andrew: South Florida and Louisiana, August 23–26, 1992.* Natural Disaster Survey Report. Silver Springs, Md.: National Weather Service.

Nunn, Kem. 1997. *The Dogs of Winter.* New York: Pocket Books.

Ocean Studies Board, National Research Council. 1992. *Oceanography in the Next Decade.* Washington, D.C.: National Academy Press.

Ocean Studies Board, National Research Council. 1999. *Global Ocean Science.* Washington D.C.: National Academy Press.

Ocean Studies Board, National Research Council. 2000. *Clean Coastal Waters: Understanding and Reducing the Effects of Nutrient Pollution.* Washington, D.C.: National Academy Press.

Orlean, Susan. 1998. *The Orchid Thief.* New York: Ballantine Books.

Paludeine, David Sean, ed. 1998. *Land of the Free.* New York: Gramercy Books.

Pell, Claiborne, with Harold Leland Goodwin. 1966. *Challenge of the Seven Seas.* New York: William Morrow.

Pilkey, Orrin H., and Katharine L. Dixon. 1996. *The Corps and the Shore.* Washington, D.C.: Island Press.

Preston, Antony. 1984. *Navies of World War 3*. New York: Military Press.

Resources Agency of California. 1995. *California's Ocean Resources: An Agenda for the Future*. Sacramento, Calif.: Resources Agency of California.

———. 1999. *California's State Classification System for Marine Managed Areas*. Sacramento, Calif.: Resources Agency of California.

Ricketts, Edward F., Jack Calvin, and Joel W. Hedgpeth. Revised by David W. Phillips. 1985. *Between Pacific Tides*. 5th edition. Stanford, Calif.: Stanford University Press.

Ridgway, Sam. 1987. *The Dolphin Doctor*. New York: Fawcett Crest.

Safina, Carl. 1997. *Song for the Blue Ocean*. New York: Henry Holt.

Sanger, Clyde. 1986. *Ordering the Oceans: The Making of the Law of the Seas*. London: Zed Books.

Savitz, Jacqueline. 1999. *Pointless Pollution*. Washington, D.C.: Coast Alliance.

Slackman, Michael. 1990. *Target: Pearl Harbor*. Honolulu: University of Hawaii Press.

Sontag, Sherry, and Christopher Drew. 1998. *Blind Man's Bluff*. New York: Public Affairs.

Steinbeck, John. 1941. *The Log from the Sea of Cortez*. New York: Penguin Books, 1986 edition.

———. 1945. *Cannery Row*. New York: Penguin Books, 1992 edition.

Stewart, Frank, ed. 1992. *A World Between Waves*. Washington, D.C.: Island Press.

Stone, Robert. 1992. *Outerbridge Reach*. New York: Ticknor & Fields.

Sullivan, Robert. 1998. *The Meadowlands*. New York: Charles Scribner's Sons.

Terrell, Bruce G. 1993. *Fathoming Our Past: Historical Contexts of the National Marine Sanctuaries*. Newport News, Va.: Mariners' Museum, NOAA.

Thorne-Miller, Boyce. 1993. *Ocean*. San Francisco, Calif.: Collins Publishers.

———. 1999. *The Living Ocean*. Washington, D.C.: Island Press.

Troll, Ray, and Brad Matsen. 1991. *Shocking Fish Tales*. Anchorage, Alaska: Northwest Books.

U.S. Congress. 1995. *Final Report on the Activities of the Merchant Marine and Fisheries Committee*. Washington, D.C.: Government Printing Office.

U.S. Coral Reef Task Force. 2000. *Coastal Uses Working Group Summary Report*. Washington, D.C.: Government Printing Office.

U.S. Department of Transportation. 1999. *An Assessment of the U.S. Marine Transportation System. A Report to Congress*. Washington, D.C.: Government Printing Office.

U.S. Federal Agencies with Ocean-Related Programs. March 1998. *Year of the Ocean Discussion Papers.* Washington, D.C.: Government Printing Office.

U.S. General Accounting Office. August 1998. *Navy Aircraft Carriers Cost-Effectiveness of Conventionally and Nuclear-Powered Carriers.* Washington, D.C.: General Accounting Office.

U.S. Senate Committee on Commerce. 1974. *The Economic Value of Ocean Resources to the United States.* Washington, D.C.: Government Printing Office.

————. 1976. *Legislative History of the Coastal Zone Management Act of 1972, As Amended in 1974 and 1976 with a Section by Section Index.* Washington, D.C.: Government Printing Office.

Wallace, Aubrey. 1994. *Green Means.* San Francisco, Calif.: KQED Books.

Warner, William W. 1976. *Beautiful Swimmers.* Boston: Little, Brown.

————. 1997. *Distant Water.* New York: Penguin Books.

Weber, Michael L., and Judith Gradwhol. 1995. *The Wealth of Oceans.* New York: Norton.

Weems, John Edward. 1980. *A Weekend in September.* College Station, Tex.: Texas A&M University Press.

Wenk, Edward, Jr. 1972. *The Politics of the Ocean.* Seattle, Wash.: University of Washington Press.

Wheelwright, Jeff. 1994. *Degrees of Disaster.* New York: Simon & Schuster.

White House. June 1966. *Effective Use of the Sea.* Washington, D.C.: Government Printing Office.

Wilcove, David. 1999. *The Condor's Shadow.* New York: W. H. Freeman and Company.

Wilder, Robert Jay. 1998. *Listening to the Sea.* Pittsburgh, Pa.: University of Pittsburgh Press.

Williams, Joy. 1997. *The Florida Keys.* New York: Random House.

Wood, Forrest G. 1973. *Marine Mammals and Man.* Washington, D.C.: Robert B. Luce.

Woodard, Colin. 2000. *Ocean's End.* New York: Basic Books.

Woods Hole Oceanographic Institute. 1995. *Ecohab: The Ecology and Oceanography of Harmful Algal Blooms.* A National Research Agenda. Woods Hole, Mass.: Woods Hole Oceanographic Institute.

Yergin, Daniel. 1992. *The Prize.* New York: Simon & Schuster.

Index

Oil Spill Prevention Act (1990), 81
oil spill trust fund, 96
oil spills, 81, 94–97, 101, 102, 223
oil tankers, 95–97
Okamoto, Mineo, 146
Oklahoma (battleship), 51
Oklahoma, 92
OPA 90. *See* Oil Pollution Act of 1990
open seas, 91
Operation Crossroads, 54
Operation Sea Dragon (1999), 64
Operation WigWam (1955), 56
Oregon, 3
Oregon State University, 55
Ostrum, Chris, 219
otter trawls, 44
Outer Continental Shelf (OCS), 18, 91,
    197, 214; fishing rights on, 175; oil
    drilling, 92, 94, 95
oxygen, 3, 152
oysters, 34, 44

Pacific Coast Federation of Fishermen's
    Associations (PCFFA), 173, 185
Pacific Northwest: shellfish poisoning, 159
Pacific Ocean: surface temperature rises,
    121; whaling, 36–37
Pacific salmon, 174
Packard, David, 25–27
Packard, Julie, 25
Packard, Nancy, 25
Packard Foundation, 25
Padre Island, Texas, 159
paints, antifouling, 54
Palm Beach, Florida, 39
Palmer, Nathaniel B., 36
Palmer Station, Antarctica, 103–109
Pamlico Sound, 158
Panama, 42–43
Panama Canal, 38, 43
paralytic shellfish poisoning, (PSP), 159
Pardo, Arvid, 13
PATH (Partnership for Advanced
    Technology in Housing), 135
Pathfinder (drill ship), 101
Paul, Alexandra, 206
Pauley, Edwin, 91–92
Pearl Harbor, Hawaii, 40; Japanese attack
    on, 43–44, 49–53

pearlwort, 107
Pecos (supply ship), 81
Pelagic Shark Foundation, 233
Pelican (research vessel), 154, 155
pelicans, 159
Pell, Claiborne, 189
Penetta, Leon, 215
penguins, 106, 107
Pennock, Jonathan (Jon), 153–154
Pennsylvania: nutrient runoff pollution,
    157
Pennsylvania State University, 54
Penobscot, 33
People for Puget Sound, 233
Pequot, 33
Perdue, James, 158
Peru, 17
pesticides, 152
Petronius platform, 88
pfiesteria, 7, 71, 156, 157, 159
pharmaceutical companies, 28–29
Philadelphia, Pennsylvania, 34
Philippines, 41
Philips, Andy, 184
phosphorus runoff, 148, 154, 157
photosynthesis, 3
Phragmites, 137
phytoplankton, 3, 4
Piccard, Auguste, 24
Piccard, Jacques, 24
Pickett, Mark, 216
Pickle Barrel Seafood, 167
Picklesimer, Glen, 78
Piianaia, Abraham, 21, 27
Pilgrims, 33
pine, 34, 35
Pittman, Walter, 57
plankton, 3
planktonic zone, 21
plate tectonics, 57
Playa Capital consortium, 134
Playa Vista, Los Angeles, California,
    134–135
Plymouth harbor, 33
poaching, 172
Point Adolphus, Alaska, 219
Point Lobos State Park, California, 248
Point Loma navy submarine base, San
    Diego, 53, 59

Wilson, Judy, 160–162
Wilson, Pete, 205, 206
wind charts, 38
Witt, James Lee, 139
Woodal, Ken, 61
Woods Hole, Massachusetts, 39
Woods Hole Marine Biological
  Laboratory (MBL), 27, 46
Woods Hole Oceanographic Institution,
  25, 46, 53, 57
World Trade Organization (WTO), 234
World War I, 43

World Wildlife Fund, 234
Wu, Norbert, 247
Wyland, 247

yellow fever, 43
Young, Don, 6, 99, 100, 165, 219
Yount, George, 86–88
Yozell, Sally, 181, 182
Yurok, 33

Zall, Linda, 67, 68
Zapata Oil, 92–93